SAFETY SECOND

SAFETY SECOND

The NRC and America's Nuclear Power Plants

THE UNION OF CONCERNED SCIENTISTS

CONTRIBUTORS

Michelle Adato, *principal author*
James MacKenzie
Robert Pollard
Ellyn Weiss

INDIANA UNIVERSITY PRESS · BLOOMINGTON AND INDIANAPOLIS

Manufactured in the United States of America

Library of Congress Cataloging-in-Publication Data

Adato, Michelle.
   Safety second.

   Bibliography: p.
   Includes index.
   1. Nuclear power plants—United States—Safety measures. 2. U.S. Nuclear Regulatory Commission. I. Union of Concerned Scientists. II. Title.
TK9152.A54   1987   363.1'79   86-45408
ISBN 0-253-35034-4

   2   3   4   5   91   90   89   88

CONTENTS

PREFACE

Founded in 1969 as an informal faculty group at the Massachusetts Institute of Technology, the Union of Concerned Scientists has grown to become one of the nation's major public policy organizations, with a staff of 35 in Cambridge, Massachusetts, and Washington, D.C., and 100,000 sponsors nationwide. UCS's activities include research, public education, and lobbying. In addition to nuclear power plant safety, UCS has worked in the fields of energy policy and national security.

The organization first became involved in the nuclear safety issue in the spring of 1971 by challenging the technical basis of the Atomic Energy Commission's performance criteria for the emergency core cooling systems of nuclear power reactors. From that time on, UCS has been looked to for sound technical assessments of the U.S. nuclear power program. Among the pioneering analyses it has published are *The Nuclear Fuel Cycle* (Cambridge: MIT Press, 1974), *The Risks of Nuclear Power Reactors* (Cambridge: UCS, 1977), and *Radioactive Waste: Politics, Technology, and Risk* (Cambridge: Ballinger, 1980). UCS has been an intervenor before, and presented expert testimony to, the U.S. Nuclear Regulatory Commission; has commented on the adequacy of proposed rules and their need and justification; and has filed petitions to the NRC to upgrade safety. Legal actions have also been successfully brought against the NRC. UCS staff members have regularly testified before congressional committees, aided several state governments, and advised foreign governments on nuclear safety issues.

Research Associate Michelle Adato is the principal author of *Safety Second*. Senior Nuclear Safety Engineer Robert Pollard contributed invaluable assistance at every stage of the report's development. General Counsel Ellyn Weiss and Senior Staff Scientist James MacKenzie reviewed drafts and offered much useful criticism. Researcher Tenley Ruth and Technical Research Associate Steven Sholly provided important research assistance. Special thanks are due to Senior Editor John Tirman, Editor Nancy Maxwell, and Legislative Counsel Michael Faden for editing. Thanks also to Jonathan Sevransky, Rhonda Kranz, Mara Williams, Vivian Myers and Nancy Jo Stockford.

Finally, UCS thanks the many friends and associates who reviewed and improved drafts of this book.

ACRONYMS AND GLOSSARY

ACRS	Advisory Committee on Reactor Safeguards
AEC	U.S. Atomic Energy Commission
ASLAB	Atomic Safety and Licensing Appeal Board
ASLAB Panel	the group of administrative judges from whom a three-member appeal board is chosen
ASLB	Atomic Safety and Licensing Board
ASLB Panel	the group of administrative judges from whom a three-member licensing board is chosen
ATWS	anticipated transient without scram
BWR	boiling water reactor
CRGR	Committee to Review Generic Requirements
DOJ	U. S. Department of Justice
ECCS	emergency core cooling systems
GAO	U. S. General Accounting Office
GDC	general design criterion
IE	Office of Inspection and Enforcement
NRC	U. S. Nuclear Regulatory Commission
NRR	Office of Nuclear Reactor Regulation
OGC	Office of General Counsel (the NRC's attorneys)
OI	Office of Investigations
OIA	Office of Inspector and Auditor
PRA	probabilistic risk assessment
SER	safety evaluation report
staff	NRC employees who report to the executive director for operations; this term does not include employees reporting directly to the commissioners
sua sponte	literally, "on its own motion"
TMI	Three Mile Island
UCS	Union of Concerned Scientists
USI	unresolved safety issue

SAFETY SECOND

ONE

INTRODUCTION

In January 1975, five commissioners were sworn in to head a new agency, the Nuclear Regulatory Commission. The NRC was a spin-off from the Atomic Energy Commission, which had been created 29 years earlier and was abolished by the 1974 Energy Reorganization Act. This historic piece of legislation recognized that the AEC had long lived with two contradictory missions: the promotion and the regulation of commercial nuclear power. The prevailing sentiment in Congress was to create a separate regulatory agency to ensure the public health and safety. The promotional activities were detailed to a new agency that eventually became the Department of Energy.

The ten-year anniversary of the NRC arrived at an especially difficult time for the nuclear industry. Dozens of plant construction projects had been canceled in the early 1980s; the cost of electricity from new nuclear power plants exceeded that from coal plants, the principal alternative; growth in electricity demand had slowed from historic highs; and many Americans believed that renewable resources and improved efficiency were preferable paths of energy development. In the wake of several industry setbacks, many newspapers and magazines declared that nuclear energy was dead. But nuclear power was not dead in one important sense. Though its future expansion may have been limited, almost 90 commercial nuclear plants were licensed to operate, with perhaps two dozen more expected to begin operations in the next few years. It will require the utmost care to ensure that these reactors pose no undue risk to the public health and safety.

It is the Nuclear Regulatory Commission's legal responsibility to ensure the safe operation of plants already on line and those not yet completed. Does the agency's decade-long performance show that it can meet this challenge? The Union of Concerned Scientists has reviewed the NRC's record in several key areas. We have found that the Commission still harbors both the promotional objectives of its predecessor and the complacent attitude identified in 1979 by the President's Commission on the Accident at Three Mile Island (the Kemeny commission).

Without major changes in the attitudes and procedures of the NRC, we are pessimistic that it can meet the challenge of regulating nuclear power in the late 1980s and 1990s.*

In this report we are, at times, very critical of the NRC's staff. We recognize that there are many technically competent staff members who are quite capable of enforcing the agency's safety regulations and who would do so if so instructed. That many do not reflects more on the NRC's upper management and on the Commission itself than on the integrity of individual staff members. Hence, while the symptoms of the NRC's poor performance are often visible in the actions of its staff, the causes of the problems must be found at the highest levels of the agency. It is here that the ultimate responsibility (and blame) rests.

This report details the lapses in regulatory vigor, the dilatory responses to safety problems, the "nothing can go wrong" attitude, the hostility to public scrutiny, and the fraternal relationship with the industry that characterized the NRC's first decade. In this introduction we first review the origins of the agency, then set the context for the specific issues covered in chapters 2 through 5.

Soon after the development of the first atomic bomb at the end of World War II, Congress transferred control over the atom from military to civilian hands with the passage of the Atomic Energy Act of 1946. With the primary purposes of maintaining U.S. nuclear superiority and protecting national security, the legislation established a five-member Atomic Energy Commission responsible for the oversight and development of every aspect of nuclear technology, both its military and its peaceful uses.

The peaceful uses of atomic energy then envisioned, primarily the production of electricity from nuclear power plants, received little attention in the AEC's early years; the agency was preoccupied with building a stockpile of nuclear weapons. In December 1953, in a speech to the United Nations General Assembly, President Eisenhower announced his Atoms for Peace plan, which promoted the peaceful uses of atomic energy under the auspices of a new U.N. agency.

Congress amended the Atomic Energy Act in 1954 to allow private

---

*In this report we have not reviewed the NRC's performance in areas other than nuclear power plant safety. However, the Commission and former Commissioner Victor Gilinsky, in particular, have often taken a positive leadership role in nuclear nonproliferation—reducing the spread of nuclear weapons abroad through misuse of the commercial nuclear fuel cycle. We applaud the NRC's efforts on this important issue.

industry to build and operate nuclear facilities. Although the amended act repeatedly refers to "the health and safety of the public," the 4,000 pages of reports, testimony, and debates discussing the legislation do not define the phrase.[1] "Nobody ever thought that safety was a problem," former AEC attorney Harold Green told an interviewer. "They assumed that if you just wrote the requirement that it be done properly, it would be done properly."[2]

In 1954, through the determined efforts of Navy Captain Hyman Rickover, the first nuclear-powered submarine, the U.S.S. *Nautilus*, was launched. In 1955 the AEC established the Power Reactor Development Program, through which large sums were allocated for nuclear development, including support for research in both industrial and federal laboratories and subsidized nuclear fuel for new plants. In an attempt to accelerate the development of commercial nuclear power, the AEC contracted with Westinghouse Electric Corporation to build the first nuclear power plant connected to a commercial grid, in Shippingport, Pennsylvania. The 60-megawatt power plant, constructed under Rickover's direction, began operations in 1957.

Despite this federal encouragement, commercial nuclear power found little support among utilities, partly because of fear of the unknown dangers of nuclear plants. To address this concern, the AEC asked the Brookhaven National Laboratory to estimate the consequences of a very large accident at a hypothetical, relatively small nuclear plant near a large city. The study, completed in 1957, estimated that up to 43,000 injuries, 3,400 deaths, and $7 billion in property damage could result from a major reactor accident.[3] To cope with the formidable barriers posed by such a possibility, Congress passed the Price-Anderson Act in 1957. (The law has been renewed twice and amended since its original passage.) This act established a $560-million limit on a utility's liability for a nuclear accident. Of this amount, only $60 million was available from private insurers; the remaining $500 million in coverage was provided by the federal government at very low premiums.*

In tandem with promotion, the AEC was charged with regulatory responsibility—in particular, making certain that the complex and untested machines were safe. The AEC began developing rules for com-

---

*As amended, the Price-Anderson Act still provides a cap on utility liability in the event of a major accident, but this limit is tied to the number of nuclear plants in operation. The limit is currently $590 million, of which $120 million is available from private insurance. The remaining $470 million would be raised through a retroactive assessment of $5 million per operating reactor.

mercial nuclear energy in 1955; these rules were intentionally vague to avoid discouraging development by applying too much red tape. "The AEC's objective in the formulation of the regulations was to minimize government control of competitive enterprise," Lewis Strauss, then chairman of the AEC, announced.[4] For example, the 1971 rule for control rooms, still in effect today, specifies only that "a control room shall be provided from which actions can be taken to operate the nuclear power unit safely under normal conditions" and that "adequate radiation protection shall be provided" for plant personnel.[5] No requirements or guidance were given at that time as to how these goals were to be achieved.

With the Price-Anderson Act, the federal government removed a major economic barrier, beginning an era of rapid growth in the number and size of plants under construction. Near the end of the 1960s, the AEC was forecasting that 1,000 plants would be operating in the United States by the year 2000. By 1969, 16 plants had actually been licensed, 54 were under construction, and 35 more were on order; five years later, the numbers were 43 licensed, 54 under construction, and 53 on order.[6]

The rapid expansion of the industry—in both number of plants and their size—laid the groundwork for many of the operating and safety troubles experienced in the years to follow. As former NRC Commissioner Peter Bradford said:

> In 1968, the largest reactor in operation was only one-half the size of the smallest reactor under construction and was one-sixth the size of the largest plant under construction. As a result, an entire generation of large plants was designed and built with no relevant operating experience—almost as if the airline industry had gone from piper cubs to jumbo jets in about 15 years.[7]

One of the first prominent public safety controversies surfaced in the early 1970s during hearings on emergency cooling core systems (ECCS), which are vital in cooling a plant's radioactive core and preventing a meltdown during certain serious accidents. The hearings, which were prompted by the Union of Concerned Scientists and which lasted almost two years, disclosed that the AEC could not validate a number of its fundamental assumptions in assuring adequate performance of the ECCS. Moreover, the Commission also attempted to suppress this bad news.[8] While hearings resulted in only minor changes to AEC criteria for acceptable ECCS performance, they underscored the problems of an agency charged with the dual mission of safety regulation and promotion. For the AEC was, first and foremost, a promoter—a role mandated by the Atomic Energy Act itself. Strict regulation was at loggerheads with the AEC's promotional bent.

The idea of breaking the AEC into two agencies, one for development and the other for licensing and regulation, had been suggested as
early as the mid-1950s, but it took almost two decades for the split to
materialize. In 1974, Senator Abraham Ribicoff, chairman of the Senate
Committee on Government Operations, which proposed what became
the Energy Reorganization Act, described Congress's concerns when he
introduced the bill:

> [T]he development of the nuclear power industry has been managed by
> the same agency responsible for regulating it. While this arrangement
> may have been necessary in the infancy of the atomic era after World War
> II, it is clearly not in the public interest to continue this special relationship
> now that the industry is well on its way to becoming among the largest
> and most hazardous in the Nation. *In fact, it is difficult now to deter-
> mine . . . where the Commission ends and the industry begins.* The result has
> been growing criticism of the safety of nuclear power reactors. . . .[9]

The proposals of Ribicoff's committee and of committees in the
House resulted in the Energy Reorganization Act of 1974, which split
the AEC into the NRC and the Energy Research and Development Administration. Congress's goal of keeping the promotional and regulatory
functions separate was unambiguous. According to the Senate report on
the reorganization act, "One of the basic purposes of this act is to separate
the regulatory function of the AEC from its developmental and promotional functions. . . ."[10] The Senate report also stated that "all references to encouraging, promoting, utilizing, developing and participating
in atomic energy or the atomic industry shall not be applicable" to the
new regulatory commission.[11]

The reorganization act gave the NRC regulatory responsibility,
sometimes shared with other federal agencies, over the medical uses of
radioactive materials, research and test reactors, nuclear waste transport
and disposal, safeguards against sabotage and weapons proliferation,
and the import or export of nuclear equipment and materials. But the
agency's primary responsibility was protecting the public health and
safety inasmuch as they might be threatened by the operation of nuclear
power plants.*

---

*The NRC is a five-member independent commission, with members appointed by
the President and confirmed by the Senate to serve staggered five-year terms. Decisions
are made by the majority of three or more commissioners present at a meeting. The
commissioners oversee administration of the agency but may delegate decisions to the staff
or to licensing boards (primarily assigned to hold hearings on NRC licensing actions). The
organization chart of the NRC as of January 1984 is given in Figure 1. A list of all the
commissioners along with the dates of their tenure is given in Table 1. The fiscal year
1985 budget for the Commission was $448.2 million. At this writing, the agency employed
approximately 3,500 persons.

TABLE 1

Tenure of Office of NRC Commissioners

William A. Anders 1975–76
Marcus A. Rowden 1975–77
Edward A. Mason 1975–77
Victor Gilinsky 1975–84*
Richard T. Kennedy 1975–80
Joseph M. Hendrie 1977–81
Peter A. Bradford 1977–82
John F. Ahearne 1978–83
Thomas Roberts 1981–90
Nunzio Palladino 1981–86
James K. Asselstine 1982–87
Frederick M. Bernthal 1983–88
Lando Zech (recess appointment) 1984–85

---

*Gilinsky was reappointed and served two terms.

The NRC's first decade was punctuated by many incidents of safety breakdowns at nuclear power plants, but one stands out as the watershed event: the 1979 accident at Three Mile Island Unit 2 (TMI-2). The accident shook public confidence in nuclear power and turned a spotlight on the inadequacy of federal regulation. President Carter ordered a blue-ribbon panel (known as the Kemeny commission after its chairman, Dartmouth College President John Kemeny) to investigate the causes and implications of the accident. While the Kemeny commission proposed a number of institutional and procedural changes, its most telling conclusion related to NRC attitudes:

> To prevent accidents as serious as Three Mile Island, fundamental changes will be necessary in the organization, procedures, and practices—and above all—in the attitudes of the Nuclear Regulatory Commission and, to the extent that the institutions we investigated are typical, of the nuclear industry.[12]

In response to the TMI accident, the NRC staff in late 1979 proposed a TMI Action Plan, which listed more than 200 actions to be taken by the NRC to upgrade reactor safety. Because of the focus of the Kemeny commission on the importance of NRC attitudes, a section on "Attitude" was drawn up. The staff opined that an independent evaluation of changes in NRC attitudes was needed to determine whether the changes were real, permanent, and sufficient. With little discussion, the commissioners rejected this proposal. The specifics of the proposal are

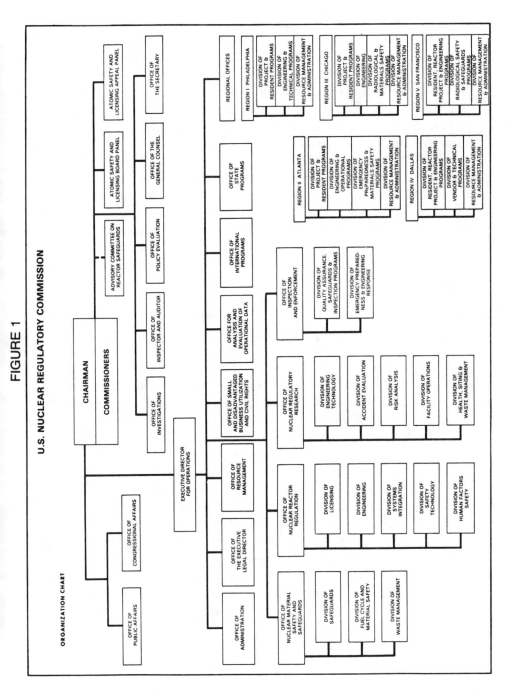

FIGURE 1

U.S. NUCLEAR REGULATORY COMMISSION

ORGANIZATION CHART

still of interest, however, because they give glimpses of the agency's mind-set and a possible answer to the question of whether it has changed enough to help prevent future accidents. The proposal contained a list of "Indicators of Inadequate Attitude," including abuse of generic issues, failure to require indicated backfits, delay in implementing backfit items once decided, failure to do realistic emergency planning, failure to believe a serious accident can happen, inadequate handling of staff dissent, siege mentality with respect to citizen intervention, allowing perceived resource limitations to thwart safety improvements, preoccupation with licensing schedules, and protecting the industry from costly changes.[13]

The TMI accident jolted the NRC into proposing a myriad of tangible safety improvements. Yet, as we shall see, the reforms precipitated by that event were insufficient and fleeting. Though some improvements in attitude were evident in 1979–80, the momentum for positive change had dissipated by 1981. Once again, the NRC's chief objectives appeared to be speedy plant licensing and uninhibited reactor operation. The desire to sweep safety problems under the rug and to ignore those who attempt to air these problems was intensifying. The TMI accident, which gave the regulators an unusual opportunity to reflect and reform, only marginally altered the operations and goals of the NRC.

In the chapters that follow, the agency's performance in four important areas is reviewed and the following questions are explored: Has the NRC tackled the toughest and most pervasive safety issues? Does the Commission accommodate public participation in its decision making? Has the agency enforced its own rules strictly and consistently? And has the NRC kept at arm's length the industry that it is directed to regulate? These are not trifling matters. Unfortunately, there is ample evidence that the answer to each of these questions is "no."

Has the NRC tackled the toughest and most pervasive safety issues? These are the generic safety issues—technical problems with reactors and plant systems that apply to virtually all reactors of the same type. We have found that the NRC has repeatedly delayed the satisfactory resolution of such issues—which are fundamental to safety—while continuing to use a "generic" label to divert troublesome technical questions from licensing proceedings. The Kemeny commission explicitly called for an end to this practice, but the NRC has failed to grapple with the technical issues themselves or the process by which such issues are diverted and delayed. Instead, the agency has played a numbers game, promoting the appearance of resolution when in fact few technical fixes are actually implemented. Indeed, the NRC has aggravated these delays by requiring specious cost-benefit analyses that have put a number of generic issues in a bottleneck.

Does the Commission accommodate public participation in its decision making? Although the public has a legal right to participate in licensing hearings, the NRC has consistently attempted to circumscribe participation through its rulings and procedures and through so-called licensing reform bills proposed to (but not accepted by) Congress. Public participation, whether by individuals, public-interest groups, or state and local governments, has long been portrayed by nuclear proponents as obstructionist. While outside involvement in licensing proceedings has been the focus of industry and NRC hostility, it has not delayed the operation of any completed plants. Where it has occurred, full and unencumbered citizen access to the licensing process has had tangible benefits. Intervenors have provided a crucial, independent check on the work of the utilities and the NRC, raising important issues overlooked or ignored by the NRC staff. In many cases public participation has resulted in improvements in reactor safety and environmental protection.

Has the agency enforced its own rules strictly and consistently? With a technology as complex and inherently dangerous as nuclear power, it is essential that safety standards be clearly drafted and consistently and firmly enforced. Unfortunately, the NRC's recent performance has been characterized by disregard for its own regulations. In the Indian Point, Grand Gulf, Diablo Canyon, and other well-publicized cases, the Commission has failed to enforce the safety standards it itself set. Such actions are deplorable, not only in these specific cases but also in the example they set for the entire regulatory regime. In the case of Indian Point, where the NRC's emergency planning rules had not been met for years, the issue was front-page news for months, and the Commission finally refused to demand adherence to its regulations. That sent a signal throughout the nuclear community that the NRC's regulations could be ignored, regardless of public outcry. The alternative is not to make the NRC a slave to formalism for its own sake. Where there is a legitimate need, the NRC has procedures for waiving regulations, provided the public health is not endangered. These formal rules and procedures were developed precisely to prevent the abuses that result from arbitrariness or favoritism, and they should not be lightly ignored or abrogated.

In the change from the AEC to the NRC, has the agency kept at arm's length the industry that it is directed to regulate? This criterion is pivotal in evaluating the NRC's first ten years. Given the unique legacy of the AEC, the NRC should have taken extraordinary measures to separate itself from the taint of industry dominance. Regrettably, it did not do so. In the NRC's licensing proceedings the agency's staff has consistently served as an advocate for the utility. As a result of inadequate NRC review, defects in design and construction have often surfaced so late in the construction process that huge cost overruns, delays, or even cancellations have become inevitable. The NRC's investigatory practices—which have included sharing drafts with those being investigated, arbitrarily restricting the scope of investigations, and undermining probes by the Department of Justice—sadly reflect the agency's lack of independence from the industry it regulates.

This review of the NRC's performance helps one understand the problems befalling the nuclear industry. No reactor has been ordered since 1978, and all reactors ordered since 1974 have been canceled. The public, the nuclear industry, and the investment community have seen the spectacle of largely completed plants abandoned when construction mistakes were caught or acknowledged so late in the game that their owners could not afford to make the necessary repairs. While the nuclear industry and the Department of Energy blame the industry's ill health on overregulation and public intervention in licensing proceedings, such claims cannot withstand scrutiny. In fact, it is misregulation and the restriction of public oversight that have contributed to the industry's troubles at Zimmer, Marble Hill, Midland, Grand Gulf, Diablo Canyon, Shoreham, Seabrook, Byron, Palo Verde, Comanche Peak, WPPSS-2, South Texas, and other nuclear plants. The agency's misregulation destroys public and investor confidence in the nuclear option, without which the industry cannot hope to recover. The result is a reduced level of safety in the nation's nuclear plants and a bleak outlook for the future of the nuclear industry.

Our judgments may at first seem harsh to those not familiar with the peculiar practices of the Nuclear Regulatory Commission. We have chosen to study particularly troublesome areas of regulatory neglect and irresponsibility because they pose the most critical threats to safety. We have also focused our study in this way because of the sorry state of "checks and balances" in government. Except for a few members of Congress and of the judiciary, the official interest in seeing that regulatory agencies are doing what they are supposed to be doing appears to be waning.

Good regulation pays its own rewards in future years. With 100 nuclear reactors likely to be operating in the 1990s, now is no time for the Nuclear Regulatory Commission to be beating a retreat from its flirtation with serious regulation in 1979–80. The industry itself will always resist aggressive enforcement of the rules, and the Reagan Administration will apparently continue to urge restrictions on public scrutiny in favor of speeding up licensing while pressuring the independent federal agencies to curtail "costly" requirements of any kind. That is not the route to a safe or efficient future. Rather, this road of regulatory minimalism is more likely to lead to increasing public opposition and accidents than to energy independence and safe nuclear power plants. In this regard, UCS repeats what it has often said: We see no inherent technical obstacles to resolving the host of safety problems plaguing today's reactors. There is serious question, however, whether the political will is present to undertake the actions needed to resolve these problems.

TWO

UNRESOLVED "GENERIC" SAFETY PROBLEMS

[C]ategorization of an issue as generic typically delays its resolution. Because issues are regarded on a general basis and are not regarded as an impediment to individual plant licensing, little incentive exists for their resolution.

The Rogovin report

Most Americans are unaware that nuclear power reactors are being allowed to operate even though scores of "generic" safety problems have not been resolved. In fact, assurances of the safety of today's nuclear reactors are belied by the existence of these problems. Generic issues, as the NRC has named them, are unresolved problems that affect either all nuclear reactors or large groups of reactors—for example, all pressurized water reactors.

The NRC maintains that generic issues need not be resolved prior to the licensing of a particular reactor, nor need they interfere with the continued operation of plants on line. The agency's rationale for this practice is that it is more efficient to address a generic issue in one consolidated proceeding than in individual licensing hearings. If the agency acted promptly to resolve generic issues, such an approach might be justifiable. Instead, the NRC has left some of the most serious safety problems unresolved for years—in some instances longer than a decade. Thus, if efficiency is the goal, it is not being served. Since its inception, the NRC has continued the AEC's tendency to delay grappling with serious reactor safety problems by labeling them generic. Indeed, the NRC in recent years has devoted an extraordinary amount of time and resources to designing procedural restrictions on the staff's ability to correct safety problems.

In this chapter, we explain the origins of the generic label and how it has been manipulated by the NRC in ways that prevent the speedy resolution of significant safety issues. Of particular importance is the way the agency has attempted to give the appearance of resolving generic safety issues and the higher-priority (though still "generic") unresolved safety issues, or USIs. Four cases are discussed, demonstrating specifically the long delays engineered by the NRC in its handling of essential problems of plant safety. Then the methods used within the agency that forestall needed improvements in operating plants are evaluated. Finally, we explore the problem of upgrading plants in order to deal successfully with severe accidents and consider what the NRC may do about this problem in the future.

The Need for the Generic Label

The push by government and industry officials in the 1960s to create a thriving nuclear power industry resulted in the licensing of reactors with numerous unresolved safety problems. Since then, knowledge of these problems has improved, but solutions have been slow in coming. The Three Mile Island accident brought many new generic problems to light and forced the government to deal with them. Yet most of the problems identified before the accident are still unresolved.[1]

The AEC and the NRC have granted construction permits to plants with the hope that resolutions to safety problems would be developed as the plants were designed and constructed.* As former NRC Chairman Marcus Rowden once noted, however, the nuclear plant operating license review comes "after a billion dollar plant has actually been built. As a result,the stakes are exceedingly high . . . which tends to skew the attitudes of the participants, if not the . . . results themselves."[2] Other NRC officials have described the same condition. The staff of the Kemeny

---

*The licensing process for a commercial nuclear power plant begins with a construction permit application and a mandatory hearing. Sometimes, as in the case of the now-canceled Clinch River Breeder Reactor, the hearing board may grant a limited work authorization (LWA) before the construction permit process is complete. The LWA usually allows site preparation, but it may allow some construction. Later, after the plant is substantially constructed, the utility files an operating license application. (Complete construction can take from six to sixteen years.) The NRC must publish a notice of opportunity for a hearing on the application. An operating license hearing is held if a person or organization successfully establishes "standing" and if the hearing board accepts at least one issue for hearing. When the operating license is approved, operation proceeds in stages: fuel loading, subcritical testing, and low-power operation.

commission, in analyzing the implications of this situation, found that defining an issue as generic was a mechanism the NRC used to "insure the granting of an operating license for an already constructed plant."[3]

In 1977, an NRC appeal board ruled that the staff must include in each plant's safety evaluation report (SER) an assessment of each relevant generic safety problem with "potentially significant public safety implications."[4] The appeal board noted that such assessments might affect its ability to make the safety findings required before issuing a construction permit. In 1978, another appeal board extended this requirement to the NRC staff's preparation of SERs for operating licenses, noting that justifying plant operation in light of unresolved safety problems would be even more difficult than issuing a construction permit under these circumstances.[5]

In 1979, when the NRC staff refined its definitions of generic safety problems, it set aside the category of "unresolved safety issues"(USIs) for issues the staff considered most important. USIs are the only generic issues reported to Congress; they are also the only generic issues discussed in the section of the SERs prepared to meet the appeal board orders requiring that generic problems be addressed before licensing. Nonetheless, some generic issues that are not classified as USIs may be just as important for safety as the USIs are.

The two appeal board orders have been followed only perfunctorily, and the NRC staff is still not required to provide adequate justification for safe reactor operation in the face of generic safety problems. The staff simply adds an appendix to its SERs discussing only the unresolved safety issues and by careful wording does not address the substance of the safety problem.

An excellent example is the staff's 1981 SER for Comanche Peak in Texas. It describes the concern over one unresolved safety issue, "Westinghouse Steam Generator Tube Integrity," as "the capability of steam generator tubes to maintain their integrity during normal operation and postulated accident conditions."* It omits the second half of the NRC's official definition of the issue, and with it the heart: "the capability of steam generator tubes to maintain their integrity during normal oper-

---

*The steam generator tubes are the only barrier preventing release of the radioactive materials in the reactor cooling system to the steam system and then the environment during normal plant operation. Following an accident, excessive leakage through the steam generator tubes could greatly increase the radiation dose to the public. The NRC has acknowledged this safety problem since at least 1975.

ation and postulated accident conditions *with adequate safety margins and the establishment of inspection and plugging criteria needed to provide assurance of such integrity.*"[6] Relying thus on a truncated misstatement of the unresolved safety issue, the SER only describes the methods "employed to minimize steam generator tube problems," discusses "the inservice inspection requirements," and concludes that the reactor "has met all *current requirements* regarding steam generator tube integrity."[7] What the SER does *not* state is that steam generator tube integrity was designated an unresolved safety issue precisely because the adequacy of "current requirements" was questionable. Neither does it discuss how the health and safety of the public is assured in the absence of a resolution of the USI, as the appeal board purportedly required.

Following the TMI-2 accident, both the President's Commission on Accident at Three Mile Island (the Kemeny commission) and the NRC Special Inquiry Group, which prepared the Rogovin report, sharply criticized the NRC's handling of generic issues. Ironically, in 1976, during the operating license review for TMI-2, the NRC's Advisory Committee on Reactor Safeguards (ACRS) recommended that "*prior to commercial operation* of TMI-2, additional means of evaluating the cause and likely course of various accidents, including those of very low probability, should be in hand to provide improved bases for timely decisions concerning possible off-site emergency measures." A report by the staff of the Kemeny commission described how the NRC staff evaded the ACRS recommendation by responding that the issue would be dealt with generically.[8] The Kemeny commission pointed to it as "an important example" of how "NRC's primary focus is on licensing and insufficient attention has been paid to the ongoing process of assuring nuclear safety." The Kemeny commission concluded that removing generic issues from the licensing process would be acceptable

> if there were a strict procedure within NRC to assure timely resolution of generic problems, either by its own research staff, or by the utility and its suppliers. However, the evidence indicates that labeling of a problem as "generic" may provide a convenient way of postponing decision on a difficult question.[9]

The Numbers Game: Sidetracking Generic Issues

Since long before the TMI accident, Congress has been attempting to move the NRC toward the resolution of generic issues, but progress has been slow. The existence of hundreds of unresolved generic issues was first brought to the attention of Congress and the public in February 1976, when NRC staff member Robert Pollard resigned, in part to protest that "the pressures to maintain schedules and to defer resolution of known safety problems prevailed over safety."[10] He cited a previously secret Technical Safety Activities Report that listed more than 200 unresolved issues being studied by the NRC.[11]

In October 1976, the NRC commissioners directed the staff to develop a plan for the resolution of generic issues. The staff came up with 355 generic issues warranting the highest priority. In 1977, Congress directed the NRC to develop a plan for specification and analysis of unresolved safety problems and to take the necessary action to implement solutions. The NRC was to report annually on its progress.[12]

The agency then engaged in a numbers game to reduce the nominal burden of generic issues without taking substantive action. The NRC's first report to Congress, in January 1978, reduced the number of unresolved generic problems to 133.[13] Its 1979 report to Congress indicated even more dramatic progress in reducing the number of unresolved problems, listing only seventeen.[14] Unfortunately, the new numbers did not reflect the actual resolution of problems. Instead, the apparent reduction was the result of new bookkeeping techniques. Some issues were lumped together, redefined, or recategorized out of the program.[15] Other issues were dismissed as not constituting a significant risk on the basis of the *Reactor Safety Study* (WASH-1400, the Rasmussen report), a risk-analysis exercise using probabilistic risk assessment (PRA), a methodology laced with uncertainties. The *Reactor Safety Study* purported to establish numerical estimates of the probability of reactor accidents. These estimates were discredited by UCS.[16] Eventually, an NRC-sponsored high-level review group also disavowed the core melt probability estimates.[17] Two months before the TMI-2 accident, the NRC officially announced that it "does not regard as reliable the Reactor Safety Study's numerical estimates of the overall risk of reactor accident."[18] At least one NRC staff member correctly observed that the "use of probabilistic risk assessment in the regulatory process is a safety issue in itself" and further suggested that the large number of safety issues dismissed "carries a cumulative level of risk that cannot be dismissed as insignificant even by WASH-1400 standards."[19] PRA is discussed further in the section "Cost-Benefit Analysis" of this chapter.

The NRC's definition of *resolved*, for purposes of removing issues from the list, was also suspect. The chairman of the NRC's Advisory Committee on Reactor Safeguards (ACRS), Max Carbon, indicated in

1979 that "in some cases an item has been resolved in an administrative sense, recognizing that technical evaluation and satisfactory implementation are yet to be completed," but that "in other instances, the resolution has been accomplished in a narrow or specific sense, recognizing that further steps are desirable, as practical, or that different aspects of the problem require further investigation."[20] In other words, some USIs were considered "resolved" when a solution had been developed, even though that solution had not been installed in the plants. Others were considered "resolved" when a conceptual solution (such as developing better means of detecting cracks in reactor vessels) had been accepted, even though the actual solution (the new crack-detection technique) was not yet developed or in use.

The NRC also changed its definition of the term *unresolved safety issue* in a way that again revealed the agency's concern that safety problems might interfere with plant licensing or operation. The staff submitted to the commissioners for review a definition that said an unresolved safety issue was a generic issue "for which it is likely that actions will be taken" to "compensate for possible major reduction in protection" of public safety or to "provide a potentially significant decrease in risk" to public safety. Subsequently, according to the Rogovin report, "concern was expressed that the definition must be compatible with the continued operation of existing plants. The commissioners thus requested the staff to revise its proposed definition."[21] The staff came up with a new definition: an issue that "poses important questions concerning the adequacy of existing safety requirements . . . and that involves conditions not likely to be acceptable over the lifetime of the plants affected."[22] That became and still is the NRC's official definition of an unresolved safety issue (USI).

The Three Mile Island accident forced the NRC to skip its annual progress report. Instead, in March 1981 the NRC sent Congress a special report identifying four new USIs, three of which stemmed directly from the accident investigations.[23] The accident disclosed more than three new safety problems, however. Indeed, the NRC had selected 425 concerns and recommendations as candidates for USIs. The agency then put these items through an intricate obstacle course composed of various sets of lengthy screening criteria. According to the report to Congress, most of the screening was "done without addressing the safety importance of an issue." For example, an issue did not pass the initial screening test if "definition of the issue requires long-term confirmatory or exploratory research."[24] The 425 concerns and recommendations eventually were whittled down to four new USIs. In 1984 the NRC reported that fifteen former USIs were "resolved" and listed twelve that were still unresolved.[25] (One additional issue had been identified as a potential USI.[26])

The Kemeny commission recommended that Congress amend the Atomic Energy Act to "require NRC to set deadlines for resolving generic issues."[27] But Congress did not act, and the NRC's "deadlines" continued to be only target dates that are routinely postponed, often for years. The Kemeny commission recommendation eventually resulted in the NRC's publication of a report in December 1983 that listed 482 generic issues (the generic issues designated as USIs were not included) and assigned 123 of them "high," "medium," "low," or "drop" rankings.[28] The NRC maintained that the remaining issues were either at or near resolution, incorporated into other issues, not related to safety, or to be given priority rankings in the future.[29] *Generic issues*, differentiated from USIs, are "possible deficiencies in the design, construction, or operation of several or a class of nuclear power plants such that the protection of the public from radiation may be inadequate."[30]

The NRC gave 29 of the generic safety issues a high priority (in addition to the USIs). Some NRC officials acknowledged, however, that there was little difference in safety importance between some USIs and some high-priority generic issues. And the General Accounting Office stated in a report in September 1984 that "at least 11 of these issues may be as important to safety" as the USIs. Thus there were more unresolved safety problems that the NRC considered most urgent than at first appeared. But, as the GAO report pointed out, only USIs "receive the focused management attention needed for an effective program."[31]

The priority rankings are used to determine which generic safety issues receive NRC resources for their resolution in future years. Although a ranking is a necessary first step, many factors undermine the reliability of the ranking; large uncertainties exist in the risk-assessment methods employed in the ranking process (including the use of plant-specific PRAs to represent entire classes of plants even though the design details of the plants differ). The GAO report noted that the rankings do not consider either risks to the environment (such as land and water-supply contamination) or multiple counting of some risks.

Perhaps most worrisome to the GAO in its 1984 report was the delay in resolution. More than four years had passed since the Kemeny commission recommended that deadlines be set for *resolving* these problems.

In 1980, the year after the TMI accident, the NRC had substantially improved its efforts to "resolve" generic safety issues (using the NRC's unique definition of the word): 97 issues had been "resolved," up from a total of nine in 1979. Most of these newly "resolved" items were safety problems identified as a result of the TMI accident,[32] and the NRC reported that the majority of these items had been implemented.[33] An average of approximately 30 issues per year had been resolved in the next three years, but most issues that had been identified in the mid-1970s remained unresolved. This pace of resolution left the agency with more unresolved issues in 1983 than at the time of the accident.[34]

Given these rates of issue resolution and new issue identification, the GAO estimated that it would take the NRC about ten years to eliminate the backlog. The GAO also pointed out that under the NRC's plans to resolve only twelve generic issues in each of the next three fiscal years, the process would take even longer. The GAO concluded that "this may be too long considering (1) the sense of urgency expressed by Three Mile Island accident review groups and (2) the age and relatively high priority of many of these issues." It recommended that the NRC take action to eliminate this backlog of issues sooner, blaming "management weaknesses" at the NRC for delays: "NRC does not . . . have sufficient management controls in place to ensure resolution of issues and implementation of appropriate changes to affected nuclear plants and to NRC's regulatory procedures in a timely manner."[35]

Unfortunately, the GAO confirmed that the NRC's definition of *resolved* still did not reflect the commonsense meaning of that word; neither the NRC nor the utilities needed to make any changes for an issue to be declared "resolved." The GAO found that the "NRC considers a generic issue resolved when 'a solution to the problem has been identified and has gone through all the necessary approval steps,' " a process that "can take from several months to 10 or more years." Furthermore, the management information system the NRC had developed to track the generic issues resolution process provided data that were incomplete and sometimes "inaccurate and misleading." The GAO concluded: "Fundamental to the resolution process should be the ability to attest to what happened to the issues after development and approval of solutions," but the NRC "does not maintain any summary information showing the disposition of resolved issues."[36]

After resolution comes implementation, another drawn-out process that can take many years. When the NRC reports that a generic issue solution has been "implemented," it again does not apply a real-world definition. As the GAO explained:

> NRC considers "implementation completed" on a multiplant action when the utility agrees to make needed changes. However, this is misleading since no actual changes in the plants have been made. As a result, the program that attempts to identify and resolve problems affecting numerous nuclear power plants breaks down at the point where improvements are to be made.[37]

The GAO also found that the NRC did not maintain any summary information showing the disposition of generic safety issues that the NRC listed as resolved. Since this information was not available, the GAO asked the NRC staff to determine how 208 generic issues had been resolved. The staff was unable to determine how 62 issues had been resolved and claimed that no changes to the plants or the agency's regulations were required to resolve 41 issues. As for the remaining 105 issues, the staff claimed that they were resolved by requiring changes to the power plants or to the NRC's regulations, but "could not differentiate between issues requiring changes to power plants or to regulations. . . ."[38] Thus, even if the NRC classifies a generic issue as resolved, it may not know how the issue was resolved; and, in cases where changes to the power plants were required, it may not know whether a particular plant made the required changes.

Despite the Kemeny and Rogovin admonitions, the NRC continued to avoid dealing with controversial issues arising from licensing proceedings by labeling them generic. In December 1981, the NRC refused to consider the effects of earthquakes on emergency response at the San Onofre plant in California, ruling that the issue would be dealt with generically.[39] By 1984, the agency had taken no action, and it again refused to let the issue be considered during licensing hearings for the Diablo Canyon plant in California by again promising generic action.[40] In July 1984, the NRC refused to require a systems interaction study at TMI-1 despite the fact that systems interactions contributed substantially to the accident at its twin reactor.* Instead, the Commission decided to "leave this matter to staff's generic program."[41] The use of the generic rubric, whether or not the problem is elevated to the status of an unresolved safety issue, clearly provides the agency another avenue of delay in dealing with crucial safety problems, regardless of the ability to reduce the numbers of unresolved problems on paper.

---

*Systems interactions involve a failure in one system inhibiting or preventing the independent functioning of another system, thereby adversely affecting safety.

Delays in Correcting Safety Problems

The NRC and the nuclear industry typically blame members of the public participating in licensing proceedings for causing "delays" in reactor operation. Not only are these charges unsubstantiated;[42] they are also ironic in light of the endless foot-dragging by the NRC and the utilities in solving important safety problems. The NRC's schedules for resolution of generic safety issues, including the highest-priority USIs, are among the most flexible to be found in any government agency. Years routinely pass between issue identification, resolution, and implementation. In several cases, accidents have occurred in operating reactors that clearly involved one or more generic safety issues. These cases include the failure of the reactor to automatically shut down at Salem in New Jersey in 1983 and the steam generator tube ruptures at Ginna in upstate New York in 1982.

One example of this drawn-out process is the handling of several significant problems identified from the TMI accident. Two years after the accident, in March 1981, three accident-related problems were finally designated as USIs.* "Task action plans" for the resolution of these issues, estimated to take from three to six months to prepare, were not issued until June, September, and December 1982—15 to 21 months later. Resolution of these issues was originally predicted to occur by April 1984,[43] but they were rescheduled and the predicted resolution dates ranged from February to April 1986.**[44] Implementation of the solutions has not been scheduled.

The histories of four unresolved safety issues—fire protection, equipment qualification, boiling water reactor pipe cracking, and anticipated transient without scram—illustrate the interminable delays in resolving some of the most critical safety problems. In certain of these cases, it took an accident—fortunately not catastrophic—to force measurable progress. It is worth reviewing each in some detail to show the NRC's approach to resolving important safety concerns.

Fire Protection

The inadequacy of fire protection in nuclear power plants was dramatically brought to light on March 22, 1975, at the Browns Ferry nu-

---

*The three USIs are shutdown decay-heat removal requirements, safety implications of control systems, and hydrogen-control measures and effects of hydrogen burns on safety equipment.

**Resolution of one issue was "not scheduled" because it was being handled on a case-by-case basis.

clear plant in Alabama when a worker checking for air leaks with a candle set fire to polyurethane and electrical cable insulation. The fire, which burned for seven and a half hours, was confined to the cable-spreading room. This room occupied a space of only 800 square feet beneath the main control room, but it was probably the worst possible location in the plant that could have burned. The fire destroyed 1,600 cables, making it impossible to monitor the status of the reactors or to control their safety devices just when information and control were most urgently needed. The destroyed cables powered almost all the redundant safety systems necessary to keep the reactor core cooled and thus prevent a meltdown.

The reactor was ultimately kept under control by a few pieces of equipment that were not even part of the plant's elaborate safety apparatus. That this equipment emerged from the fire undamaged was random chance. Engineers of the Tennessee Valley Authority, operator of Browns Ferry, stated privately that a potentially catastrophic radiation release was avoided "by sheer luck."[45]

NRC Chairman William Anders said that "the fire revealed some deficiencies in design and administrative controls at Browns Ferry. . . ."[46] Any implication that the AEC and the NRC were unaware of these deficiencies until they were disclosed by the fire is incorrect. The record of regulatory action preceding the fire shows that when the AEC licensed Browns Ferry it knew of the serious safety deficiencies that caused the fire and its severe impact.[47]

Twelve days after the fire, the NRC issued "a broadened directive to operators of all nuclear power plants to review procedures for orderly shutdown and cooldown of the reactor should normal and preferred alternative systems be inoperative." Licensees were also instructed "to review and, if necessary, revise procedures and policies for use and control of combustible materials, ignition sources, and fire fighting equipment during construction, plant modification, or maintenance."[48] The NRC set no firm schedule for this plant-by-plant review.

As a result of NRC directives, inspectors warned supervisors to monitor their plant workers and told utilities to write a letter to the NRC stating that they had met the new requirements. Plant workers ran fire drills, on-site fire brigades assumed greater importance, and inspectors made more frequent tours. More fire extinguishers were provided, fire hoses were installed that could reach all parts of the plant, hydrants were multiplied, and so on. Yet any action to go beyond these minor fixes to test for materials flammability, to redesign and rebuild cable-spreading

rooms so the electrical cables would be physically and functionally independent, to install independent means of shutting down the plant in
the event emergency power is lost—in other words, to change the conditions that caused the damage at Browns Ferry—was indefinitely delayed.

Almost a year after the accident, a special NRC group studying the
fire concluded that it had "revealed some significant inadequacies in
design and procedures related to fires." The group's report pointed to
"the inoperability of redundant equipment for core and plant cooldown
[that] shows the present separation and isolation requirements should
be reexamined."[49] This subject, the separation and isolation of safety-
related circuits, became the center of a major controversy. It was the
lack of such isolation at Browns Ferry that allowed the fire to disable
what were supposed to be independent and redundant safety systems.
The most prudent corrective actions—which the NRC had recognized
long before but did not require—would have been either to isolate safety-
related circuits from their backup circuits or to build separate cable-
spreading rooms.[50]

The inadequacy of the NRC's response to the fire was brought to
the attention of Congress in February and March 1976 during hearings
on reactor safety. The hearings were held after Robert Pollard resigned
his position as a project manager for the NRC and Dale Bridenbaugh,
Richard Hubbard, and Gregory Minor resigned their high-level positions
in the nuclear power division of General Electric Company. During these
hearings, the NRC said several things that were contradicted by subsequent events.

One telling example concerned an NRC testing program being performed by Sandia National Laboratories "to provide experimental data
to evaluate the adequacy of the separation criteria" specified in the NRC's
fire-protection guidance. The criteria at that time specified that redundant circuits must use flame-retardant cable and be separated horizon-
tally by three feet or vertically by five feet. The NRC told Congress that
the "results of these tests are expected to confirm the adequacy and
conservatism of the separation distances now serving as guidelines for
current staff review."[51] Contrary to the staff's assurances to Congress,
the results of the tests conducted at Sandia Laboratories in July 1977
failed to confirm the adequacy of the NRC standards. Electrical cables
meeting the flame-retardancy standards had been arranged to simulate
the two redundant sets of safety cables in a nuclear plant. The fire,
deliberately started in one set of cables, "propagated across a five foot

vertical separation simulating a division having redundant safety cable."[52] In November 1977, as it became clear that the NRC was not going to act on this new evidence, UCS petitioned the Commission to amend its regulations so that nuclear plants would be adequately protected from fire. UCS asserted that the Sandia tests showed that under current standards for physical separation of redundant cables, flame retardancy, and cable tray loading, a fire could cause unacceptable damage to cables powering crucial redundant safety equipment and thus leave the plant defenseless against an accident. In UCS's view, that constituted undue risk to the public.[53]

The NRC's initial response to the UCS petition contradicted its 1976 congressional testimony predicting that the Sandia tests would "confirm the adequacy" of the existing standards. The staff now maintained that the Sandia tests "confirm the need" for measures that "go well beyond a simple reliance on existing standards for electrical cable separation and fire retardancy of redundant cables." Despite this strong wording, the staff concluded that "actions taken and underway . . . provide adequate protection to the public health and safety pending full implementation of the current criteria."[54] The Commission did not explain how actions not yet implemented could provide adequate public protection.

In response to the UCS petition, the Commission accepted the need for new fire-protection regulations. In February 1981—more than three and one-half years after the first UCS petition and almost six years after the Browns Ferry fire—a new fire-protection rule took effect.[55] In contrast to the NRC's 1976 assurances to Congress that its separation criterion of three feet between redundant safety circuits was adequate, the equivalent portion of the new rule required twenty feet between these circuits. The belated fire-protection regulations appeared impressive, but in UCS's judgment they still did not result in adequate fire protection. The reason was twofold: The rule itself was rife with legal and technical loopholes, and the NRC was demonstrably reluctant to enforce it.

Although the rule was entitled "Fire Protection Program for Nuclear Power Facilities Operating Prior to January 1, 1979," plants licensed to operate before that date needed to meet only three of the rule's fifteen standards. In addition, the technical loopholes were so large that a plant could meet all the fire-protection regulations and still have no means of preventing a major accident. For example, the rule was based on the assumptions that a fire would not occur at the same time as any other accident and would not be a consequence of any other accident. In fact, the potential for fire is particularly high during some types of accidents.

A hydrogen fire or explosion inside the reactor building, for example, is only possible following a severe loss-of-coolant accident. And it is following an accident that safety-system equipment, such as large pump motors, draws the most electrical current—precisely the condition when a loose connection is most likely to initiate a fire.

Another loophole in the rule permitted all equipment necessary to bring the plant to cold shutdown (below approximately 200° F) to be vulnerable to a fire if at least one set of equipment could be repaired within 72 hours. Having to rely on the hope that fire damage can be predicted well enough to ensure prompt repair of equipment gives little room for comfort. The rule also permitted all equipment except one set needed to bring the plant to hot shutdown (subcritical, but with the temperature greater than approximately 200° F) to be damaged by a single fire. That means that if any equipment other than that damaged by the fire should fail to operate properly or if the reactor operators make mistakes, there may be no way to bring the plant to a safe condition. The long sequence of equipment and human failures that characterized the TMI-2 accident attests to the likelihood of multiple failures. In addition, the NRC routinely allows plants to operate when some safety equipment is inoperable or down for maintenance. There is thus no basis to believe that at least one system will remain operable after a fire.

Additional loopholes existed in the new rule's liberal provisions for granting exemptions and the long schedules permitted for installing the necessary protection features. In a fourteen-month period, the NRC granted 234 technical exemptions to various parts of its fire-protection regulations, as well as numerous schedule exemptions.[56] Between the legal and the technical loopholes that pervaded the fire-protection regulations, many plants still lacked the fire barriers and automatic fire-suppression systems necessary to protect redundant safety circuits. More than ten years after the Browns Ferry accident, a fire could still destroy the plant systems needed to prevent a large radioactive release to the environment.

Equipment Qualification

The NRC's handling of the equipment qualification issue is illustrative of its failure to resolve a safety issue promptly—indeed, at times, to apparently obstruct safety progress. The term *equipment qualification* (usually referred to as environmental qualification) is shorthand for the process of testing and analyzing safety equipment to ensure that it can function in environmental conditions caused by an accident—steam, elevated temperature and pressure, and radiation. A regulation requiring important safety equipment to be qualified, General Design Criterion 4, has been on the agency's books since 1971. A 1978 ruling by the Commission, in a case brought by UCS, acknowledged that equipment qualification is *"fundamental* to NRC regulation of nuclear power reactors."[57]

UCS first brought the equipment qualification issue before the Commission in a petition in November 1977 after concluding that results from NRC-sponsored equipment qualification tests demonstrated the inadequacy of the NRC's standards. In the tests, electrical equipment meeting NRC standards was subjected to simulated accident conditions. The components failed. The day after UCS raised this issue, the NRC staff issued a statement saying that UCS had misconstrued the safety significance of the test results and that none of the equipment that failed the tests was used in any plant's safety systems.[58] Both claims were later shown to be false.

Shortly thereafter, the NRC nonetheless asked operators of nuclear plants to supply information detailing the status of equipment qualification at their plants. For two and a half years, the utilities largely ignored the NRC requests. Then, in 1980, the Commission adopted strict new standards for equipment qualification, stating that the prior standards were clearly inadequate. It gave the utilities almost two more years—until June 30, 1982—to meet the new rules.[59] The Commission directed that each operating plant's license be amended to incorporate the 1982 deadline to ensure that the rules would be enforced. Such an action is rare. No licensee requested a hearing to oppose this license amendment.

A year later, in June 1981, a group of utilities asked the Commission to postpone the deadline one year because they could not meet it. Without an extension, they argued, they would have to shut down their plants. The NRC staff responded to this request with three proposals, including one more generous than the utilities' request: a two-year extension. The staff also added exceptions that could potentially have allowed unlimited delay.

In arguing for an extension of the deadline, the staff informed the Commission that more than 80 percent of the safety equipment in operating plants still had not been shown to be capable of functioning properly when exposed to the harsh environment caused by an accident. Nevertheless, the staff estimated that a substantially smaller proportion of the electrical equipment—15 to 40 percent—was "unqualified" and would have to be replaced. This estimate was based on the staff's evaluation of about 20 percent of the essential safety equipment.[60] The staff argued, in effect, that the reactors should be allowed to remain in operation until the NRC determined which safety equipment was defective. On the day of the 1982 deadline, the Commission issued an "immediately effective" rule waiving the deadline altogether but providing no opportunity for public comment or hearings on this action.

On October 8, 1982—five years after the Commission (in response to the UCS petition) began its efforts to get licensees to demonstrate the qualification of safety equipment and four years after the Commission's ruling that equipment qualification is "fundamental to NRC regulation of nuclear power reactors"—the staff gave the Commission a report.[61] It showed that the staff's estimate that 15 to 40 percent of the electrical equipment was unqualified had been optimistic. The staff now reported that 44.6 percent of the electrical equipment in operating plants had to be replaced, physically modified, relocated, shielded, or further tested and that 31.1 percent lacked documentation to determine whether it was environmentally qualified or needed to be replaced, modified, relocated, shielded, or tested. Altogether, 85.8 percent of the electrical equipment relied upon to protect the public in an accident was not shown to be environmentally qualified, and only 6.6 percent was fully qualified.* (Just as disturbing as the report itself was the fact that the NRC kept this information from the public until October 1983, when it was released in response to a UCS Freedom of Information Act request.)

The Commission apparently did not recognize how sharply the report contradicted assurances the staff had given in the spring of 1982 before the Commission suspended the June 1982 deadline. In UCS's view, the impression then created by the staff was that most electrical equipment was qualified and that adequate justification for continued safe operation had been obtained for the equipment that was not shown to be qualified. Unfortunately, the staff neglected to tell the Commission that neither the staff nor its contractor, Franklin Research Center, had independently verified the licensees' claims concerning equipment listed as qualified. Indeed, as Commissioner James Asselstine subsequently said:

> except in a very limited number of cases, the staff has not performed the detailed examination of supporting documentation needed to verify in-

*7.6 percent of the equipment did not need to be qualified because it did not provide a safety function or would not, because of its location, be exposed to a harsh accident environment.

dependently that either the equipment is properly qualified or that there is an adequate justification for continued operation. Nor is there any reason to believe that an in-depth examination of all the licensee's documentation will occur any time in the near future. In the very few cases where staff has begun such in-depth reviews the evidence indicates that licensee efforts have been inadequate.[62]

On June 30, 1983, a U.S. Circuit Court of Appeals ruled that the NRC had violated the Atomic Energy Act, the Administrative Procedure Act, and its own rules when it suspended the deadline without providing an opportunity for public comment.[63] NRC lawyers had defended the agency's action by arguing to the court that the NRC rule requiring an opportunity for public comment "in the manner stated in the [Federal Register] notice" had been met because offering *no* opportunity to comment was one way of regulating the *manner* of commenting.[64] The Commission's lawyers had also argued that the NRC's central justification for the deadline waiver—that continued operation of the plants despite noncompliance with equipment qualification standards would not unduly risk public health and safety—was an insignificant "ancillary finding" or "explanatory background remark" that was "not a part of the rule."[65] However, without such a finding the Commission could not lawfully waive the deadline contained in each operating license.

After the court's ruling, the NRC's resistance to adhering to the law continued. Not until eight months after the court ordered the Commission to "provide an opportunity for comment on the sufficiency of current documentation purporting to justify continued operation pending completion of environmental qualification of safety-related equipment" was it provided. It is UCS's view, however, that the Commission still did not comply with the court's decision because it failed to reach the required finding that individual plants were sufficiently safe despite their operation with unqualified safety equipment. The Commission asserted that the 1982 deadline was not a "substantive safety standard" but was only meant to "urge licensee compliance with the environmental qualification program."[66] Moreover, technical information that the staff used in judging the status of qualification at individual plants—necessary for writing informed comments—was not made available to the public, in spite of a Freedom of Information Act request filed by UCS.

In September 1984, the Commission voted once again to revoke its June 1982 deadline. As a concession to Commissioner Asselstine, the only member to vote against the revocation, the NRC required each utility to certify that it could meet plant-specific qualification programs by the new deadline of March 1985 or the second refueling outage after March 1982, whichever came first. (Reactors are shut down to refuel generally once every twelve to eighteen months.) If a utility could not so certify, it had to apply for an exemption. There was no guarantee, however, that these utility certifications would be any more reliable than those submitted in the past. It was also apparent that many plants would not be able to meet the new deadline; at least seventeen utilities requested extensions. In dissent, Asselstine summarized the status of the NRC's seven-year effort to bring the industry into compliance:

> Given the evidence of continuing instances of licensee recalcitrance and the hurried and superficial reviews of the plant-specific deficiencies identified by the commenters on the proposed rule, there is no basis in this rulemaking record for concluding that all licensees are now pursuing effective environmental qualification programs with due diligence.[67]

Boiling Water Reactor Pipe Cracks

Cracking has been observed in boiling water reactor (BWR) pipes since 1960.[68] One cause of recent pipe cracks, intergranular stress corrosion cracking (IGSCC), is believed to occur over time through the combination of oxygen in cooling water, carbon in the steel pipes, and stress that may have been caused by earlier welding. It generally appears after about two years of plant operation. Cracks have been found in reactor cooling-system piping and parts of emergency core cooling systems, and can be numerous; for example, since 1983 more than 50 cracks have been found in Browns Ferry Unit 1.[69]

The hairline-thin cracks begin near welds inside the inch-thick pipes and thus may not be visible to the eye until they have penetrated the pipe wall. The safety concern is that the cracking could cause a pipe break without warning. Some affected pipes are as large as 28 inches in diameter. If a pipe were to break, cooling water would be lost; if emergency systems failed, a meltdown of the reactor core could result.

A 1975 internal NRC working paper described this cracking as an unresolved safety problem that would be resolved in 1976.[70] No such resolution occurred. Little was done until March 1982, when plant operators visually detected a through-the-wall crack in a pipe at Nine Mile Point 1 in upstate New York. (Much of the piping has since been replaced.) After this discovery, partial inspections (whose ultrasonic testing methods later proved inadequate) were required at some BWR's. On July 14, 1983, Harold Denton, director of the NRC's Office of Nuclear Reactor Regulation and an infrequent advocate of plant shutdowns, told the Commission of his concern about five plants and announced that they would have to shut down for inspection within 30 days. The next day, industry representatives met with the commissioners and convinced them to overrule Denton's decision. Explaining the reversal, Richard Vollmer, a top-level NRC engineer, said that "industry gave them a good enough story that said [safety] . . . was outweighed by the costs of down-time to the utilities. That's what it came down to."[71]

The Commission has allowed BWRs with cracks smaller than a speci-fied size to go on operating largely because of continuing inspections, short-term fixes, and the assumption that cracked pipes will leak before they break, allowing time for the plant to be shut down and repaired. Nevertheless, the Advisory Committee on Reactor Safeguards ques-tioned the reliability of this "leak-before-break" theory in August 1983. The ACRS also described reliance on one of the ultrasonic techniques commonly used to measure crack depth as a "delusion, since we can find no consistent experimental evidence or body of expert opinion indicating that the measured crack depths bear any direct relationship to the actual crack depths."[72] The ultrasonic testing inspection methods, using sound waves to detect the cracking, have proved so undependable that the extent of cracking in any given plant still cannot be known, even if all the welds are checked, which is not required in all cases. For example, the through-the-wall crack discovered visually at Nine Mile Point had not been detected by ultrasonic testing nine months earlier.[73] At Peach Bottom 3 in Pennsylvania, ultrasonic testing indicated a crack depth only halfway though the wall when the pipe actually was cracked all the way through.[74] Once found, most cracks have been sized by teams that later failed crack-sizing tests. (The industry is slowly retraining inspectors.) The NRC has estimated that full implementation by all plants of the countermeasures, including improved inspections, will take several years.[75]

In July 1984, eight years after the NRC had planned to resolve the problem, the staff reported that it had confirmed the causes. But the staff did "not feel that IGSCC represents such an urgent problem that it necessitates immediate additional regulatory action." While a long-term solution was proposed, inspections would continue for now. For newer plants, where cracking had not yet occurred, a preventive thermal treat-ment was planned, along with changes in water chemistry.[76] For plants where cracking was the most widespread, the NRC recommended, but did not require, pipe replacement with low-carbon steel. The staff report said that pipe cracking had been confirmed in 19 of the nation's 30

BWRs.[77] (Most of the others had not operated long enough to develop the problem.) A few months later, in September 1984, only five plants had replaced or were replacing piping, and two others were slated for replacement in about a year.[78] The agency's reluctance to require pipe replacement can be traced to the tens of millions of dollars it could cost for each reactor. Vollmer, the NRC engineer, indicated in an interview in October 1984 that what the agency views as a relatively low risk combined with very high cost is a factor in the NRC's reluctance.[79]

Anticipated Transient without Scram

One of the most serious of the unresolved safety issues, anticipated transient without scram (ATWS), remained unresolved for almost a decade and a half until an accident—which the NRC staff called "the most serious precursor" to a core meltdown since the TMI accident[80]—forced the agency's hand.

When a reactor has an "anticipated transient"—a routine malfunction, such as a loss of normal electrical power or a turbine shutdown that requires shutdown of the reactor—control rods are supposed to insert automatically into the reactor core to halt the fission process, an action referred to as a scram. If the rods fail to insert or are delayed even briefly, the accident is called an anticipated transient without scram, or ATWS. An ATWS event could result in rupture of the reactor vessel or piping due to overpressure, with a possible release of large amounts of radioactivity.

ATWS has been officially recognized as a generic safety problem since at least 1969. As early as 1973, the AEC staff and the ACRS agreed that an additional diverse reactor shutdown system should be required for new plants.[81] The industry was successful in forestalling any new requirement, claiming that the probability of an ATWS event was exceedingly remote. The accident probabilities contained in the Rasmussen report, while recognized by the agency to be unreliable, were cited as support for this argument. Ironically, the report identifies ATWS as one of the most significant contributors to core-melt probability.[82]

In 1978, the NRC staff again found that the potential for an ATWS event presented an unacceptably high risk to the public and thus rejected the industry's proposal to dismiss the problem on probabilistic grounds. But, because no acceptable design for a redundant shutdown system had been proposed, the staff concluded that other "mitigating" systems were the "now promising alternatives."[83] In February 1979, the NRC asked the industry for information, with answers due by April 1979. In 1980, the NRC concluded that "the information submitted by the industry fell far short" of the request.[84] Finally, in November 1981—twelve years after the problem was officially recognized—the NRC proposed an ATWS rule. The industry again opposed any new requirements. Another year passed, and the NRC set up a task force that drafted recommendations, presenting them to the NRC's new Committee to Review Generic Requirements (CRGR). The committee held at least three inconclusive meetings on ATWS in 1982 and early 1983.

Then, on February 22 and February 25, 1983, the Salem plant in New Jersey had two total failures of the automatic scram system—ATWS events. Fortunately, the reactor was operating at only 20 percent and 12 percent of full power when these events occurred, and the operators quickly scrammed the reactor manually.[85] Had the reactor been at full power, prompt operator action—within 90 seconds—would have been required to prevent severe damage to the reactor core.[86]

The result? The NRC set up the Salem ATWS Generic Implications Task Force, which met with the Committee to Review Generic Requirements four times and briefed the Commission at least twice. In total, the CRGR held eight meetings on ATWS. The Commission finally published an ATWS rule in June 1984,[87] but implementation was expected to take a number of years. Furthermore, the additional equipment required by the rule does not have to meet all the requirements that normally apply to safety-related equipment, such as seismic qualification, physical separation, and redundancy.[88]

Institutional Bottlenecks

In recent years, the Commission's focus appears to have shifted further from resolving safety issues and more toward devising institutional means of restricting the issuance of new requirements. (The NRC has also proposed a program that would screen *existing regulatory requirements* for their "effectiveness" and then decide whether they should be eliminated. The worthiness of these requirements would be evaluated by comparing their contribution to risk reduction, judged through a combination of qualitative judgment and PRA, with the cost to the industry and the NRC of complying with them. In proposing this program, the staff recognized that the agency might encounter "strong resistance" to eliminating existing requirements. It noted that even the industry might have reservations "because of the potential destabilizing effect of controversy over proposed regulatory changes and uncertainties concerning the outcome of the effort."[89] The Commission has been preoccupied with the "backfitting" issue, which is defined by the NRC as the establishment of new requirements on plants after the construction permit or operating license has been issued. This preoccupation was the apparent result of the industry's often-repeated complaint that the NRC issues new requirements at the drop of a hat. It is a dubious claim in light of the glacial speed of the agency's action in addressing serious generic safety problems of long standing. In fact, NRC Chairman Nunzio Palladino told Senator George Mitchell in September 1983 that the Commission could offer no examples of unnecessary backfits.[90] Similarly, despite repeated requests from Congress, the industry has been unable to document a basis for the claim that the NRC imposes requirements that are unnecessary. If there have been some cases where backfits were of questionable safety value, they are the exception rather than the rule.

Nonetheless, the agency has spent an extraordinary amount of time and resources responding to industry allegations of a backfitting problem. While the concept of improving the process to ensure that new requirements are well considered and justified is certainly supportable, NRC actions have aimed less toward this goal and more toward creating a system in which licensees can avoid or substantially delay the implementation of needed safety improvements. These actions demonstrate a dangerous retreat from the "safety first" standard of the Atomic Energy Act.

The Backfitting Rule

The NRC's backfitting rule, 10 CRF 50.109, requires that in order to issue a backfit, the NRC must find that the action will "provide substantial additional protection which is required for the public health and safety." Industry leaders maintained that the rule had not been followed and that the NRC staff frequently issued backfits with no prior analysis. Rather than making sure that the current rule would be followed in the future, the NRC, at the industry's urging, began working on a new rule in 1982 to cover plant-specific (as opposed to generic) backfits.

In addition, in 1983 the NRC instituted new "interim" backfitting procedures (to be used until a new backfit rule is passed) that require a number of layers of staff review, a multistage appeal process for the utility, and, if the utility is still dissatisfied, extensive cost-benefit analysis prior to issuing a new requirement.[91] While this process gives utilities multiple chances to argue against the new requirement, it offers no opportunity whatsoever for public participation in the decision making. Nevertheless, with the exception of Commissioner Asselstine, the Commission wanted a permanent backfitting standard that would place even stricter limitations on the issuing of new requirements.

The Regulatory Reform Task Force, composed of NRC staff members, was charged with drafting the new backfitting rule. The task force proposed a rule with highly restrictive standards for the imposition of backfits, and the Commission promulgated the rule in September 1985. These standards must be met even when a utility has no objection to the backfit, which Edson Case, the NRC's deputy director for regulation, said will "put an inordinate burden on the staff and detract from the safety responses."[92] Even before requesting information from a utility—for example, about the safety status of a piece of equipment—the NRC staff will have to do an evaluation of the "burden to be imposed" by the request and ensure that it is "justified" in view of the safety significance.[93] Such restrictions on the NRC's ability to ask questions could cripple its safety research and enforcement programs.

The rule requires the staff to automatically prepare a cost-benefit analysis for every backfit. It also specifies a number of highly prescriptive standards to be followed, including consideration of the potential reduction in risk to the public versus installation and the downtime or construction-delay costs to the utility. The staff has to show that the backfit will result in a "substantial increase in the overall protection" to the public and that the "direct and indirect cost of implementation" of the backfit is "justified in view of this increased protection." Asselstine, in opposing the rule, cited a September 1984 NRC report: "Operating experience indicates that a total loss of a safety system is not a rare event," that there are "component performance and reliability problems," and that "Emergency Safeguard Systems are frequently challenged."[94] He stated that in light of unresolved safety issues and new safety issues still being identified as a result of operating experience, the NRC must retain the ability require changes that would reduce the risk of reactors to as low as reasonably achievable. Asselstine said that the "emphasis in the rule on cost considerations, particularly when coupled with the high standard for imposing a new requirement, is likely to have a strong chilling effect on the staff's consideration and development of new safety requirements."[95]

Cost-Benefit Analysis

Cost-benefit analysis appears reasonable on its face; hardly anyone is opposed to preventing wasteful expenditures. Since new requirements can be expensive and safety improvements must to a certain degree be prioritized, it is perhaps difficult for the NRC to avoid some cost considerations. However, the benefits of improvements are extremely hard to quantify. The NRC has a specific mandate, deriving from the Atomic Energy Act, to make safety the paramount consideration in its decisions. The consequences of a serious accident are too great to allow cost considerations to intrude on necessary safety improvements. Unfortunately, such an intrusion appears to be taking place.

Several problems, both practical and legal, arise from the NRC's use of cost-benefit analysis. Perhaps most significant, the problems in performing meaningful cost-benefit comparisons are overwhelming. In fact, while NRC backfitting procedures and proposals are presented as "rationalizing" the process, the present-day use of cost-benefit analysis actually makes the process irrational. The most fundamental problem is that the benefits—the reduction of accident risk—cannot be fairly quantified.

The degree of risk reduction attributable to a safety improvement is first estimated and then translated into dollars by the NRC in a chain of highly complex and controversial calculations, of which the following is a greatly simplified description. Probabilistic risk assessment (PRA) is used to derive two estimates of the probability of an accident—one for the plant as is, the other assuming the safety improvement is made. Similarly, PRA is used to estimate one of the consequences: the radiation dose to the public, expressed in "person-rems,"[*] for the two cases. These radiation exposures are then multiplied by the associated accident probability in order to convert them to an average radiation dose per year of plant operation. Finally, the two average radiation-exposure figures are converted to dollars by multiplying them by $1,000, the arbitrary value assigned to avoiding one person-rem of exposure. The "benefit" value of the safety improvement is, according to this technique, the difference between the figures for the as-is plant and the backfitted plant. If the difference is larger than the cost of the backfit, the safety improvement is deemed "cost-beneficial."

---

[*]A person-rem is a unit of population radiation dose. For example, one person-rem would result from one person receiving one rem of exposure, or ten people receiving one-tenth of a rem exposure each.

At each point in the calculations used to translate risk-avoidance into dollars, great uncertainties are not given due consideration. Complex quantitative "fault-tree/event-tree" analyses are used to determine accident probabilities and consequences, but these analyses are based on thousands of potentially arbitrary and largely unverified assumptions and thus can be manipulated to reach a predetermined result. Because of their great complexity and cost, PRAs are totally inaccessible to the public and cannot be independently scrutinized.* In addition, the bottom-line results of any PRA are enormously imprecise; the NRC has conceded that for the most serious accidents, the uncertainty bands can range to a factor of 100 or more.[96]

The Advisory Committee on Reactor Safequards has been critical of the NRC's use of PRA. In June 1982, the ACRS warned the Commission against undue reliance on it:

> The large uncertainties inherent in PRA are well recognized. . . . These uncertainties make the use of PRA in safety decision making (which occurs already within the NRC) subject to large differences in the results obtained by different groups of analysts for the same accident scenario. These uncertainties also permit abuse of the methodology to obtain a result which supports a predetermined position by selective choice of data and assumptions.[97]

In September 1982, ACRS members Myer Bender and Jeremiah Ray had even harsher words for this methodology:

> PRA studies as currently performed will remain inscrutable and will, at least for the next decade, be little more than a display of logical thought based on essentially arbitrary assumptions.

> The claims for PRA concerning its ability to assess public safety risk are little more than a sham that will hide the fact that the basis for safety will always depend upon the judgment of a few individuals.[98]

The PRA cannot be relied on for sound decision making. There is nothing magical about a PRA analysis that allows a decision maker to proceed directly to the "bottom line"—in this case, to a decision on whether a proposed backfit is "cost-effective." Rather, properly used, PRA can be a valuable supplement to the decision-making process by highlighting vulnerabilities in plant design and operation. PRA is also useful in illuminating where and how uncertainties might affect the decision maker's ultimate conclusion. Thus PRA can be a valuable adjunct

---

*The NRC spends about $500,000 simply to *review* a PRA; performing one requires $2 to $10 million or more, depending on the scope of the analysis.

to the decision-making process, but it cannot reliably provide "yes" or "no" answers except in the most extreme examples—in which case the answer was probably obvious in the first place without spending millions of dollars on the PRA analysis.

In addition, NRC's cost-benefit analyses ignore whole classes of benefits. The full economic benefits of avoiding a serious accident, which could range to the billions of dollars, are not incorporated. (Offsite insurance claims for the Three Mile Island accident totaled $31.25 million as of March 1983, of which $5 million was used to set up a TMI public-health fund, according to the NRC. GPU Nuclear, the licensee, estimated that the cleanup of TMI-2 would cost more than $1 billion.[99]) As Commissioner Asselstine pointed out in criticizing the Commission's proposed backfitting rule:

> [T]he list of factors to be considered under the rule in evaluating the benefits and costs of proposed backfits heavily weights the evaluation in favor of cost considerations. Potential benefits such as the reduction in the likelihood of serious operating events and accidents, avoiding the loss of onsite property, and assuring the long-term availability of the plant's generating capacity are ignored, while every possible cost, including such questionable "costs" as the NRC resource burden, is emphasized.[100]

Moreover, there is much evidence to suggest that certain NRC-proposed (and current) uses of cost-benefit analysis may be illegal. At a meeting in May 1984, an NRC staff member told the Commission that the "limitation in our ability to consider cost is not a new one" and read an exchange that had taken place in hearings held on the Energy Reorganization Act of 1974. Senator Abraham Ribicoff had asked L. Manning Muntzing, then AEC director of regulation: "Does the AEC ever allow cost to stand in the way of installing the newest safety safeguard devises?" Muntzing had replied: "[U]nder the Atomic Energy Act, the AEC is charged with assuring the reasonable safety of the facilities and licensees, and for that reason, the first decision is made with regard to safety. If it costs additional money, it costs additional money."[101]

NRC case law, articulated first during the Maine Yankee operating license hearings and maintained thereafter, has held that the Atomic Energy Act does not allow costs to be considered in making safety decisions:

> in the safety sphere the evaluation of the risks attendant to reactor operation is not undertaken as an element of a [National Environmental Policy Act]-type process by which costs may be traded off against benefits. Rather, the function of the evaluation is to ascertain whether the ultimate,

unconditional standards of the Atomic Energy Act and the regulations have been met; e.g., whether the public health and safety will be adequately protected.[102]

The Atomic Energy Act calls for the Commission to issue safety regulations that define the minimum necessary for adequate protection of the public health and safety, and these regulations then must be met. Cost should therefore not be considered in determining whether changes that are necessary for compliance with these standards should be made. Similarly, costs should not be considered if the proposed backfit involves a new method of implementing the Commission's regulations when previously approved implementation methods were based on mistakes of fact and as a result there is no longer a finding of compliance with the regulations. (The NRC's Office of General Counsel (OGC) made these two points to the Commission, stating that the Atomic Energy Act prohibited the consideration of costs in these situations when considering the backfitting of plants with construction permits.[103] Yet the NRC's proposed backfitting rule may require consideration of costs in these situations. For plants with operating licenses, the OGC said costs may be considered in determining compliance *schedules*.[104])

That is a crucial distinction. The great majority of required changes result not from adopting new regulations but from the discovery, too often after the reactor has already been licensed to operate, that it does not in fact meet the regulations it supposedly met when it was licensed. Most of the NRC's rules are very vague. General design criterion (GDC) 3, for example, requires simply that a reactor be adequately protected against fire, while GDC 4 requires only that equipment be qualified to withstand an accident environment. Neither contains any specific requirements. Over the years, as events, such as the Browns Ferry fire, and research program results led the NRC to discover that licensed plants did not provide the degree of protection supposedly assured by the regulations, more detailed guidance was issued to see that they were met. The great majority of the post-TMI accident fixes also fall into this category.

Robert Minogue, the NRC's director of research, explained in an internal memordandum how this situation came about, as reactor designs went from 60 to 1,000 megawatts in little more than a decade:

> This rapid increase in power levels occurred without obtaining any substantial operating experience on the nuclear plants at intermediate levels prior to increasing the size to higher levels. . . .

I think that much of the [backfitting] of regulatory requirements for the operating plants about which industry has complained has been a direct result of the fact that the unduly rapid push to larger sizes has resulted in what amounts to a generation of prototypes . . . this situation was almost inevitable given the substantial extrapolation from the early technology.[105]

Consideration of costs is also prohibited when decisions are made during the operating license stage on issues that were left unresolved at the construction permit stage. The U.S. Supreme Court ruled in 1961 that safety issues may be left unresolved at the construction permit stage only with the understanding that "the Commission is absolutely denied any authority to consider" money spent during construction in its decision on whether to allow operation of the plant.[106] The Office of General Counsel confirmed this understanding to the Commission in a memo in May 1984.[107]

At a meeting the same month, NRC staff member John Montgomery, of the Office of Policy Evaluation, explained to the Commission that the limits on considering costs in dealing with issues left unresolved at the construction permit stage are a result of a "bargain" struck early on with the industry:

[The] system was developed deliberately as an accommodation to the nuclear power industry. It was developed in order to get plants licensed without resolving significant safety questions as in the [Power Reactor Development Company] case. In effect, a kind of deal was struck very early in the industry. The Commission said okay, go license plants without insisting on a great deal of detail in the construction permit application and we'll let you build plants without delaying processing of the application while we resolve significant safety questions but in return we expect you to accommodate us in making changes or allowing backfits later on without regard to cost.[108]

Significantly, even when the OGC concludes that cost may legitimately be considered, "cost cannot be accorded equal or greater weight than safety in any safety-cost tradeoff."[109]

Finally, the Supreme Court has ruled in an analogous case that unless a statute specifically provides that safety considerations should be weighed against cost, federal agencies are not authorized to do so.[110] The Atomic Energy Act contains no such provision.

Committee to Review Generic Requirements

In the face of a clear mission to make safety the paramount consideration in decisions concerning new requirements, the NRC has allowed cost to stand in the way of the swift implementation of necessary safety fixes. This action has occurred through the cost-benefit analyses required by the Committee to Review Generic Requirements, and it may occur more extensively if the proposed backfitting rule is adopted. The CRGR, headed by Victor Stello, Jr., was established in late 1981 to "control" new requirements. At that time, industry complaints about backfitting had reached a crescendo following the issuance of post-TMI "lessons learned" requirements. According to the CRGR's charter, one objective of this control is to "remove any unnecessary burdens placed on licensees." All staff proposals for generic requirements must be reviewed by the committee, which then is to make recommendations as to whether they should be adopted. The charter states that the CRGR is to make sure that new requirements will contribute "effectively and significantly" to public health and safety and will use NRC resources in "as optimal a fashion as possible."[111] Unfortunately, the CRGR in practice appears to function largely as a bottleneck.

If a decision-making process assessing the need for safety improvements is not accessible to all parties in the debate, the public cannot trust it to consider all relevant factors. The CRGR is almost completely insulated from public scrutiny. Only the NRC staff and consultants may attend its meetings; it allows only limited access to relevant documents and eventually publishes only brief minutes of meetings. In 1983, the NRC adopted a policy of withholding CRGR meeting minutes from the public until final resolution of the issue is achieved, claiming that until then they are "predecisional." There are no transcripts by which the public can ascertain how the CRGR arrives at its decisions.

The best picture one can get from such a restricted view of the committee is that of a black hole, into which issues fall and never emerge. The CRGR sends many items back to the staff repeatedly for more review or modifications. It requires extensive, time-consuming cost-benefit analyses (generally using potentially inflated industry cost estimates without serious scrutiny) and demands new analyses if it is not satisfied with the results of the first ones. The result is a logjam from which items sent back to the staff sometimes never resurface, with no date in sight for the next CRGR review, much less a deadline for final agency resolution. While Chairman Palladino has commended the CRGR for having "sharpened the thinking in the agency on whether new staff proposals really had a good safety payoff,"[112] Commissioner Victor Gilinsky observed that "the committee functions as a fairly undiscriminating bottleneck.

The effect of this is to hold up everything, including mere information notices of safety problems. I think we need a different approach here."[113]

The committee's interest in cost-benefit analysis is clearly not balanced by an equally strong interest in resolving safety problems. Three cases in which the CRGR appears to have caused long delays are worth highlighting. Two of them involve unresolved safety issues.

In the first case, Proposed Regulatory Guide SC78–4, "Qualification and Acceptance Tests for Snubbers Used in Systems Important to Safety," the safety concern is the uncertain adequacy of "snubbers" used to hold pipes in place during events that would cause excessive vibration and could result in pipe breaks. The CRGR held two meetings on this proposed regulatory guide in March and May 1982. It asked the NRC's Office of Nuclear Regulatory Research to analyze a number of aspects of the benefits of this guide in light of an expensive estimate of implementation costs supplied by utilities.[114] The research office was still working on the required information in October 1984—nearly two and a half years later—with the next CRGR review not scheduled until spring of 1985.[115]

The second case involves USI A–43, "Containment Emergency Sump Performance." Following a loss-of-coolant accident in a pressurized water reactor, water discharged from the leak or pipe break collects on the containment building floor and in the containment emergency sump. The concern behind this USI is that successful long-term cooling of the reactor depends on whether the sump provides adequate, debris-free water to the emergency core cooling pumps. The CRGR supported two proposals by the Office of Nuclear Reactor Regulation (NRR) that would, according to the CRGR, "reduce requirements on future [operating license] applicants." But it questioned NRR's cost-benefit analysis for another proposal requiring plant-specific analyses to assess the potential for debris blockage of sumps in operating plants.[116] At two meetings, in November and December 1982, the CRGR maintained that NRR's analysis of this requirement overstated the benefits (which NRR had concluded outweighed the costs) and understated the costs. At the second meeting, the committee told NRR to "review their risk reduction analysis in light of the analysis performed by [CRGR] with the objective of developing the most realistic assessment" of the benefits of the proposal.[117] The package went out for public comment, and NRR performed a more sophisticated analysis that was not reviewed by the CRGR until July 1984, more than a year and a half later. At that meeting, NRR told the CRGR that the new analysis again led to the conclusion that utilities should perform plant-specific analyses to assure that they were adequately protected. NRR noted that the "proposed backfit would not provide substantial additional protection over that thought to exist, but would assure a level of safety previously thought to exist"[118]—the minimum required by the Commission's regulations and the Atomic Energy Act. The CRGR, however, decided that NRR's new analysis was too conservative and that "resolution of CRGR concerns" would prove that the benefits do not justify requiring licensees to "expend the resources to demonstrate that their designs are acceptable." The CRGR said it would reconsider the proposal or a modified one in another meeting after NRR again responded to its concerns.[119] At this writing, no further CRGR meeting had been scheduled.

The third case involves USI A–9, "Anticipated Transient Without Scram." As already described, this safety problem has been on the AEC/NRC's unresolved list for more than a decade. In November 1982, the CRGR directed the NRC staff to prepare an extensive cost-benefit analysis to support the proposed ATWS rule. The Office of Nuclear Regulatory Research performed this analysis before the CRGR's second meeting on the rule in January 1983. At that meeting, the CRGR decided that it was unable to determine the total safety benefits and costs of implementing all the modifications identified by the staff, and the research office agreed that the ATWS rulemaking package would attempt to address them all. At the next meeting, after the Salem plant suffered two ATWS incidents, the CRGR suggested that a plant might reach an age when the benefits of full implementation might not justify the costs. In May 1983, four meetings later, the CRGR was still requesting additional cost-benefit justification.[120]

In February 1983, the NRC staff demonstrated a by-the-numbers approach to safety in reporting what it seemed to consider an accomplishment:

> There has been a substantial reduction in the number of new generic requirements imposed on reactor licensees since the inception of CRGR. For instance, the staff had projected there would be 1900 new operating reactor licensing actions generated in FY–1982, whereas the actual number was less than 900 new actions. Most of this reduction was in the number of multi-plant actions and can, I believe, be attributed to the existence of CRGR. Thus, the CRGR had a major impact in the staff's achievement of reducing the operating reactor licensing backlog from about 5400 to 3600 actions during FY–1982.[121]

The Neglect of Severe Accidents

A paramount attitude that the Kemeny commission identified as contributing to the TMI-2 accident was the NRC's failure to believe that a serious accident can happen. Similarly, the NRC's post-TMI Lessons Learned Task Force noted that the "mindset" problem regarding serious accidents was "probably the single most important human factor with which this industry and the NRC has to contend."[122]

The TMI-2 accident caught the plant's operators and the NRC off guard because it went well beyond what the plant was designed and the operators were trained to cope with. The accident involved a series of equipment and human failures officially deemed to be incredible prior to their occurrence. Since then, the issue of the response of nuclear power plants to severe accident conditions has been a key question before the NRC. Shortly after the accident, the concept of a "Degraded Core Rulemaking" was proposed to determine how and whether the agency should adjust its regulations to contend with such accidents.

In the years that followed, the NRC strayed from the rulemaking course and eventually decided that a policy statement would suffice, thus limiting the public's legal means for having a say in the decision. The proposed severe accident policy statement, which at this writing was before the Commission for approval, concludes that as a general matter all currently licensed plants are adequately safe and that severe accident issues no longer need to be considered in the absence of a startling new development. The statement also mandates a cost-benefit standard that could prove a nearly insurmountable barrier to the staff's ability to order backfits.[123]

The premise of the policy statement, that current plants are adequately safe, runs contrary to both the state of the art in probabilistic risk assessment and the state of knowledge regarding severe accident phenomena. The statement relies on only a limited number of PRAs to represent all plants, although each plant is unique. It states only that at some undetermined time in the future, very limited plant-specific analyses will be required. Commissioner Asselstine asked that more comprehensive analyses be conducted at all plants (as suggested by the Advisory Committee on Reactor Safeguards), that a board be established to review staff and industry recommendations on certain issues, and that differing views be heard. His suggestion got a cold reception from his colleagues.[124] Asselstine maintained that the policy statement does not support the conclusion that operating plants are adequately safe: "The policy . . . does not even define the specific technical questions. My question is why aren't [the staff] and the industry willing to put it on the table?"[125]

The staff was visibly anxious for the Commission to adopt the severe accident policy statement. At a meeting in October 1984, the head of the NRC staff, Executive Director for Operations William Dircks, argued that this policy represented a "major milestone in achieving a stable and predictable licensing process."[126] In his eyes, such a policy was "long overdue" because "[t]here are decisions out there waiting to be made . . . [these] investment decisions deserve some policy guidance."[127]

While deliberations continued, the existence of a draft of the policy statement affected how the agency did business. Work on generic safety problems was in danger of being halted. At a meeting in April 1984, the CRGR discussed the need for a "temporary moratorium" on imposition of new requirements for any "big ticket" generic items, including some of the highest-priority unresolved safety issues. Such a moratorium would last until the severe accident policy statement and a "safety goal" policy were put in effect and the NRC's work on revising accident source terms was further along. (The NRC proposed the adoption of "safety goals" for nuclear power plants. Decisions to adopt safety improvements to reduce risks below the goals would be based on cost-benefit calculations. The incremental cost of improvement would not be more than an arbitrarily assigned $1,000 for each person-rem avoided as a result of the improvement.[128] An NRC staff report on accident source terms—estimates of radioactive (and nonradioactive) releases from plants as the result of accidents—was scheduled for completion in early 1985. It was expected to propose lower source terms—and thus reduced accident-consequence estimates—than those in use. The agency may use these revised source terms to justify regulatory changes, such as relaxing emergency planning and reactor siting standards.[129]) The reason for the moratorium was to be certain to derive the most "cost-effective" requirements.[130]

The CRGR noted the difficulties a moratorium would cause, including its "conflicts or inconsistencies" with "current NRC commitments to schedules for resolution of USIs" and the perception that the NRC "may be crossing a threshold between economic regulation and safety regulation." The committee concluded, however, that crossing this threshold was probably fine because "protection of the utility investment is normally consistent with protection of the public health and safety."[131] Unfortunately, protection of the public is not necessarily consistent with the CRGR's quest for cost effectiveness and the reduction of short-term expenditures.

The proposed severe accident policy statement says that "the Commission intends to reserve for its own deliberations the resolution of severe accident issues affecting plants under construction." Therefore, the Commission has stated, these issues cannot be considered in public hearings.[132] Before the TMI-2 accident, the agency dismissed degraded-core accidents as incredible events; since then it has considered severe accident issues in only a very limited fashion. Thus the NRC appears to have returned to its pre-TMI-2 mindset.

THREE

THE PUBLIC AS ADVERSARY

> It is absolutely amazing, the lengths to
> which the Commission will go to avoid find-
> ing that a party is entitled to a hearing on an
> issue.
>
> Commissioner James K. Asselstine

When the Kemeny commission concluded that the NRC paid "insufficient attention" to the "ongoing process of assuring nuclear safety," it
attributed this negligence largely to the fact that the agency's "primary
focus is on licensing."[1] The Kemeny commission pointed to the delayed
resolution of generic safety issues as a significant consequence of this
limited focus, but the NRC has expressed its preoccupation in many
other ways. Perhaps equally prevalent is the Commission's negative attitude toward public participation in licensing proceedings and toward
those who exercise the right to participate by becoming "intervenors"—
formal, legal participants—in licensing cases.

The goal of the large majority of intervenors—and their legal right,
as those most directly affected by NRC decisions—is not to cause delay
but to scrutinize the safety of a potentially highly hazardous reactor being
sited in their midst. Since hearings generally begin long before construction is completed, they need not cause delay. In fact, in September 1983
the Commission told Congress that public hearings had never delayed
the operation of a single reactor.[2] More important, the list of contributions made by members of the public to improvements in plant safety
and environmental protection is a long one and, together with the principles of democratic decision making, weighs heavily in favor of maintaining full public participation.

Despite the benefits to be gained from public participation in the
licensing process, the NRC has, with rare exceptions, regarded intervenors as adversaries. As Commissioner Peter Bradford told Congress
(after the Commission had ruled that utilities could challenge the ne-

cessity of post-TMI requirements but intervenors could not challenge their adequacy):

> The Commission is poorly informed as to the history, motivations and consequences of citizen groups' intervenors. It seems to be basing its decisions instead on some hypothetical concept of a vampire intervenor with whose imagined potential transgressions it is obsessed to the point of curtailing all interventions to avoid a few abuses.
>
> The citizens whose right or privileges of inquiry are at stake here are not some crew of anti-nuclear crazies unless the Commission succeeds in driving three-fourths of the nation into that category. . . . They are people raised on the American notion that they have a legitimate right to inquire into events affecting the fundamental nature of their communities and they will not relinquish that right to salve the nightmares of three federal officials.[3]

(After the UCS filed suit to challenge this policy, the Commission withdrew and replaced it with a more equitable rule allowing debate on both the necessity for and sufficiency of the post-TMI "fixes."[4])

As this chapter shows, the NRC's attitude toward the public is demonstrated by the behavior of the agency's staff and hearing boards in licensing hearings, by specific Commission actions unfavorable to public participants, and by the NRC's attempts to "reform" the licensing process by restricting public participation. Contrary to the agency's intentions, the result has often been to cause licensing delays when safety problems that had been overlooked or ignored had to be remedied. Another result, perhaps more significant to the long-term health of the nuclear industry, has been to undermine confidence in the industry as problems and delays mounted and as the public failed to see the NRC as an impartial regulator with the public interest at heart.

The Licensing Process

When the Atomic Energy Act of 1954 was passed and amended in 1957, a trade-off was made. Nuclear power was exempted from state and local regulation by federal preemption of on-site reactor safety issues, the nuclear industry's liability was limited, and insurance was provided by the Price-Anderson Act. In exchange, Congress mandated full, open public hearings to resolve safety issues before each reactor could be licensed. [5]

The Atomic Energy Act requires that licensing hearings be held before the NRC can issue a construction permit, regardless of whether members of the public choose to participate as parties to the hearing. But hearings are not automatic later, after the plant is under construction and prior to the granting of an operating license; members of the public who wish to raise safety or environmental issues at that point must request a hearing and meet the applicable standards for admission as "intervenors."*

To preside over hearings, the NRC has an Atomic Safety and Licensing Board Panel composed of administrative judges who are usually assigned in groups of three, each group being an ASLB, or licensing board. To review these boards' decisions, the NRC has an Atomic Safety and Licensing Appeal Board Panel, also composed of administrative judges assigned in threes; each of these groups is an ASLAB, or appeal board. Most of these judges are not lawyers; it is typical for an ASLB to have two engineers or academics, with a lawyer as chair of the panel. Any party to a hearing may appeal specific issues in a licensing board's decision, but an appeal board's review takes place even if no party appeals. The Commission may elect to review appeal board decisions. Final recourse is to the federal courts. However, a licensing board's decision becomes "immediately effective" unless stayed by an appeal board, the Commission, or the courts. In other words, if a licensing board's decision is favorable to the utility, plant construction or operation generally continues during the appeal process. (After the TMI accident, the Commission amended its rules to require that the commissioners decide the "immediate effectiveness" of construction permit approvals and decisions authorizing operation above 5 percent power.[6] The Commission is considering reverting to the old rule.)

Unfortunately, although the machinery for public participation exists, powerful forces subvert its effectiveness. Both a general, institutional opposition within the NRC and specific instances of direct intervention by the Commission act to undermine public participation.

The Reality of the Licensing Process: A Weighted System

In general, the public is made to feel that it is an interloper in NRC proceedings, caught between a licensing board that the Commission has told to move cases quickly, a utility that is eager to get rid of intervenors, and the NRC staff, which often simply seconds the utility's motion. Ex-

---

*State and local governments can also intervene in licensing proceedings or may participate as an "interested state" without taking a position on any issue. NRC hearing procedures are contained in 10 CFR Part 2.

perienced attorneys involved in their first NRC case have commented
that they have never seen a system so weighted against the presentation
of intervenors' points of view. Indeed, a principal finding of the NRC's
post-TMI Special Inquiry Group in the Rogovin report was that "insofar
as the licensing process is supposed to provide a publicly accessible forum
for the resolution of all safety issues relevant to the construction and
operation of a nuclear plant, it is a sham."[7]

The NRC's licensing process presents a number of institutional barriers to effective public participation. To begin with, participation utilizing lawyers, consultants, expert witnesses, and depositions can easily
cost $50,000 or more. Most intervenors are hard pressed to afford even
minimal participation and cannot hope to compete with the resources
of the utilities and the NRC in presenting their case.

Moreover, the structure of the licensing process puts intervenors at
a disadvantage. The process actually begins when an applicant for a
construction permit submits the first version of its preliminary safety
analysis report (PSAR) and environmental information to the NRC staff
for an initial "completeness" review. This review does not deal with issues
of substance. The substantive staff review and the documents that em-
body it—the safety evaluation report (SER) and the draft environmental
statement (DES)—are months from publication. Even the utility's basic
licensing documents are in preliminary form; it is typical in the months
before, during, and after the hearings for the utility to issue a dozen
and sometimes as many as 50 or more amendments containing the detailed information required to show compliance with NRC rules.

Nevertheless, the licensing "clock" starts when public notice of the
construction permit application is given, even though the NRC's formal
review is in progress and the case is not nearly ready for hearing. That
has two unfortunate effects. First, intervenors' attempts to gather the
information necessary to file contentions and to prepare for the hearing
are treated as "delay." (Contention is the term used by the NRC for an
allegation. Contentions constitute an intervenor's statements of reasons
why the permit or license should not be issued. A factual basis must be
provided for each contention.[8]) Ironically, intervenors must formulate
their positions before the NRC staff's own review is completed. Second,
the licensing boards often appear to view intervenors as impediments to
achieving the goal of a speedy hearing. Because detailed safety information is not available until relatively late in the process or is modified
after contentions have been formulated and discovery completed, intervenors may be required to move to amend contentions, reopen dis-

covery, or modify the hearing schedule in order to fairly address the relevant issues. Such motions not only require intervenors to meet a special burden; they also place the intervenors in the position of causing "delay."

By the time the NRC staff completes its safety review, it has as a practical matter informally resolved any major differences it may have had with the utility. From that point on, the staff's advocacy of the utility's position places these two parties to the hearing in an adversary position toward intervenors.* This fact explains what intervenors have learned by hard experience: No matter how technically credible the intervenors are or what legitimate issues they raise, the staff makes little attempt to meet with them, to consider seriously whether their technical concerns are valid, or to determine what if any corrective action should be taken. Instead, the staff's immediate response, with rare exceptions, is to find some justification for opposing the intervenors' positions on all substantive and procedural issues, a stance that continues during the entire hearing process.

Throughout the process, it appears to be the goal of the utility, and generally of the NRC staff, to keep potentially troublesome information off the record. To this end, rules are frequently interpreted narrowly to exclude unfavorable evidence and disregard serious safety issues. In the 1983 San Onofre operating license proceeding, for example, the licensing board (conforming to a Commission ruling on the same case) ruled that intervenors could not present evidence showing that hospitals around the plant would not be adequate to treat persons injured in a large accident. It held that NRC rules only require the applicant to *list* existing facilities and that therefore "boards are not to go beyond lists of existing facilities to determine whether those facilities are adequate to cope with various accidents."[9]

Similarly, in TMI-1 restart hearings held in 1980–81, the Union of Concerned Scientists raised the issue of whether the safety equipment at TMI-1 could function under such harsh accident conditions as high heat, pressure, and radiation. The NRC staff objected to allowing into evidence its own safety evaluation report that assessed the compliance or lack thereof by the operator, General Public Utilities, with NRC requirements for just such "environmental qualification." The report showed many deficiencies. The staff claimed that it was "outside the scope

---

*See the section in chapter 5 on "Licensing Hearings: The NRC Staff as the Applicant's Advocate."

of the hearing," and the licensing board refused to receive the staff's report into evidence.[10] Later, in writing its final decision on technical issues at TMI, the licensing board admitted that without the evidence in the rejected staff report it had "no basis" for determining the status of equipment qualification or the risk of operation with the unqualified equipment. The board conceded that it was "unfortunate that the staff objected to the receipt into evidence" of this document. Nevertheless, rather than reopening the record to admit the document and subjecting its authors to questioning, the board merely directed the staff to "certify" to the Commission the status of equipment qualification.[11] Such certification is simply a staff report sent to the Commission; no questioning concerning the basis for the staff's conclusions is permitted. Ironically, after arguing that the NRC report which UCS had attempted to introduce into evidence was irrelevant, the staff certified the status of equipment qualification by sending this very report to the Commission. (As late as March 1984, an NRC staff audit of TMI-1 equipment qualification files for eight component types revealed that the documentation to demonstrate equipment qualification was not even close to adequate in any of the files audited.[12])

In some cases, the NRC has deferred consideration of certain safety issues brought to its attention by the public during the construction period on the grounds that such issues should await the operating license review. Then, once the billions of dollars required to build the plant have been spent, unbiased consideration of many safety issues becomes extremely difficult.

In one instance, intervenors* in the Seabrook construction permit proceeding raised the issue that the thousands of people who used the beaches within a few miles of the plant could not be evacuated during an accident—an issue directly related to the site's suitability. The appeal board in 1977 interpreted the NRC's rules as establishing that an accident requiring evacuation could never occur; therefore, the feasibility or infeasibility of evacuation was irrelevant to licensing the plant.[13] Shortly after the TMI-2 accident, rules were promulgated requiring emergency planning for a ten-mile zone around all plants, prompted by the recognition that serious accidents can happen at even the safest plants.[14] The Seabrook plant was not yet a *fait accompli*; construction of Unit 1 was approximately 25 percent complete and Unit 2 was barely started.

---

*The Seacoast Anti-Pollution League and the New England Coalition Against Nuclear Pollution.

The intervenors quickly petitioned the NRC to immediately consider whether adequate emergency protection could be provided to the public surrounding the Seabrook site. They urged the NRC to hold hearings on this issue before the plant was built, knowing full well that it would be too late afterward. The NRC refused to allow hearings, maintaining that the utility's risk of loss of investment could not be a factor in later decision making and that this issue could be fully explored during the operating license proceeding.[15] The intervenors, including the Commonwealth of Massachusetts and most of the towns surrounding Seabrook, pursued the issue in the operating license hearing. It was, however, far too late. By that time the public's choices were to accept either the direct risk to their safety or the financial cost of abandoning the plant.

Finally, there is an NRC doctrine that when the Commission amends a license to add some new requirement, the utility may claim a hearing to challenge the *need* for the requirement but the public is precluded from challenging its *sufficiency* to cure the safety problem in question. The doctrine is followed even when it was the public that first brought the safety problem to the NRC's attention. Thus the utility and the NRC staff can privately negotiate a resolution that may be inadequate and the public has no opportunity to present evidence challenging this resolution. While this practice was upheld by the D.C. Circuit Court of Appeals,[16] it remains unfair and detrimental to safety.

"A Neurotic Set of Appetites"—Intervention by the Commission

A practice that the Commission has used with increased frequency since 1981 to curtail public hearings is to step uninvited into ongoing hearings and halt inquiry into safety issues. This practice has included reversing licensing board rulings and halting the consideration of issues that licensing boards deemed so serious that the boards raised them independently or pursued them when intervenors ran out of resources (referred to as *sua sponte* review). *Sua sponte* reviews by NRC licensing boards have been rare, undertaken only when a board believes an issue is of such importance that the public interest requires the NRC to address it whether or not an intervenor has raised it. An NRC licensing board discussed the importance of *sua sponte* review, noting that "if serious safety matters are potentially involved in an operating license hearing, the public interest would not condone a Licensing Board having its hands tied by being wholly dependent upon an intervenor" to litigate a case.[17] The board cited a U.S. Court of Appeals decision:

> In this case, as in many others, the [Federal Power] Commission has claimed to be the representative of the public interest. This role does not permit it to act as an umpire blandly calling balls and strikes for adversaries appearing before it; the right of the public must receive active and affirmative protection at the hands of the Commission.[18]

For this reason, NRC rules allow licensing and appeal boards to examine issues independently of intervenors when they determine that "a serious safety, environmental, or common defense and security matter exists."[19] Four cases illustrate the NRC's tendency to remove the authority from its licensing boards to consider evidence on safety issues: San Onofre, Comanche Peak, Zimmer, and Indian Point.

San Onofre. In 1981, the licensing board for the San Onofre operating license proceeding raised, *sua sponte*, the issue of the impacts of earthquakes on emergency planning—asking, in effect, whether the utility and the NRC staff had considered the possibility that an earthquake that damaged the reactor might simultaneously disrupt evacuation routes and sever offsite communications.[20] The plant is in particularly earthquake-prone southern California.

The Commission issued an order in December 1981 directing the board to drop consideration of the issue. Instead, the Commission promised that it would be pursued "generically." Commissioner Victor Gilinsky, in dissenting from the order, wrote that "it appears the Commission will go to any length to avoid having a Licensing Board deal with a question the Board itself has raised." He observed that a commonsense approach would be to allow a board to examine the issue and make a decision according to the unique circumstances of the case; the Commission could review the board's decision later if it saw fit. Gilinsky also predicted the likely effect of making the issue generic:

> If past practice is a guide: Interagency meetings will be held. Memoranda will be written. The Commission will be briefed. Contracts to study the question will be awarded to national laboratories. Increased budget requests will be received from our staff. The Commission will be drawn into ponderous rulemaking. But the most elementary steps to assure public protection will not be taken. An all too familiar story.[21]

In commenting on the Commission decision in this case, Commissioner Bradford described the lack of evenhandedness the Commission had displayed in choosing in which cases to interfere:

> When [the Commission] has stepped into proceedings in progress, it has curtailed investigation of issues unfavorable to the applicant; the Commission has stayed its hand when that action upholds Board or staff conduct favorable to the applicant. It has rarely required a Board or the staff to expand safety or environmental consideration. . . .

> Despite a recent demonstration of the value of *sua sponte* review, the Commission is telling a Board that has had the foresight to uncover "a serious safety matter" . . . that it may not inquire into the matter further, even though the Board apparently doubts that is has "reasonable assurance that adequate protective measures can and will be taken in the event of a radiological emergency (10 CFR 50.47)." The result of this action could easily be an inadequacy in San Onofre emergency planning that goes unremedied for a long time.[22]

Bradford's examples of the Commission's actions that upheld board or staff positions favorable to the utility included: Three Mile Island, where the Commission restricted the manner of litigation of hydrogen-control issues;[23] Bailly, in Indiana, where the Commission refused a hearing on a change in plant foundations,[24] a decision later reversed by a Circuit Court of Appeals;[25] South Texas, where the Commission would not review an appeal board order reversing a licensing board decision to furnish names of witnesses interviewed during an NRC investigation;[26] and Diablo Canyon, where the Commission would not review a licensing board decision authorizing a low-power license.[27] In the Diablo Canyon case, the Commission suspended the license two months later when a serious quality-assurance breakdown was discovered.

Time proved that Gilinsky was overly optimistic in his prediction of what would follow from the Commission's decision to address earthquake effects on emergency plans "generically." Three years after blocking the board's attempt to resolve the issue at San Onofre, without having the generic review, the Commission repeated its action at Diablo Canyon, another plant sited in California, three miles from an earthquake fault.

Commissioner James Asselstine, in dissenting from the majority's 1984 decision to grant Diablo Canyon a full-power license, criticized the Commission's reliance on the purported low probability that an earthquake would disrupt emergency response and its attempt to again call the issue generic. He argued that the Commission had previously recognized that Diablo Canyon was located in an area of uniquely high seismicity and that to rule out the chance that an earthquake could affect emergency planning simply defied common sense.* He criticized the Commission's refusal to face this safety concern:

> In its apparent determination to avoid adjudicating an issue that the agency itself has acknowledged to be material to emergency planning, the Commission has repeatedly changed its mind about how to treat this issue only to end up right back where it started three years ago—promising a generic rulemaking.[28]

---

*For a more extensive discussion of the Commission's treatment of this issue at Diablo Canyon, see "Ignoring Earthquakes and Regulations at Diablo Canyon" in chapter 4.

In December 1984, the Commission majority proposed a generic "resolution," stating that earthquakes need not be considered in emergency planning anywhere.[29]

Comanche Peak. Three weeks after the Commission refused to allow the San Onofre licensing board to raise the issue of the effects of earthquakes on emergency planning, it stifled another licensing board's independent efforts. This time, at Comanche Peak, in Texas, the boards concerns were over eight safety issues raised by an intervenor group* that was forced to drop out of the proceeding because it could no longer afford the "massive" costs involved. Among these issues were whether NRC regulations were met regarding fire protection, protection against rupture of the reactor vessel, and the ability of safety systems to function during an accident. The utility and the NRC staff urged the board to dismiss the intervenor's issues, but the board felt that there was insufficient information available at that time to draw conclusions on these issues that "may have significant health and safety consequences," and chose to consider them *sua sponte*.[30]

The Commission ordered the board to drop the issues. Commissioner Bradford again observed that the Commission had recently refused to correct grave errors in board decisions favoring utilities, but here:

> a rodent, errant *sua sponte* has been spotted in our fields and for this the agency eagle unsheaths its talons and plunges from the clouds. A more neurotic set of appetites than the ones leading to this plunge is hard to imagine. . . .
>
> This action shows the agency's Supreme Court [the commissioners] to share the dullwitted licensing process priorities of the American Nuclear Energy Council [an industry lobby group].[31]

Zimmer. In July 1982, the licensing board in the Zimmer (Ohio) operating license proceeding decided to review, *sua sponte*, eight issues raised by the intervenor (Miami Valley Power Project) relating to quality assurance at the plant and the corporate character and competence of the utility, Cincinnati Gas & Electric Company. In raising the issues, the intervenor had not met the NRC's standards for timeliness and reopening of the record, but the board believed that the public interest nonetheless required reopening the record.[32] In an unusual stance, the NRC staff supported the board's review because the issues were so serious.[33] The board noted:

> The staff has identified Zimmer as a plant with a serious quality assurance breakdown.

---

*ACORN, Association of Communities Organized for Reform Now.

[W]e believe that a full public airing of this matter will not only contribute to public confidence, but will also strengthen the [quality-assurance] program. Subjecting this program to the scrutiny of the Commission's adjudicatory process can only contribute, not detract, to reasonable assurance that the public health and safety will be protected.[34]

Nonetheless, the Commission ordered the board to halt its review. While it acknowledged that serious issues existed, it claimed that the staff was already investigating these matters.[35] Ignoring both the licensing board and the staff's advice, the Commission decided that the quality-assurance breakdown would be treated outside of the adjudicatory process.* Commissioner Asselstine issued a strongly worded dissent:

For years, we have heard the charge that the Commission's regulatory process denies members of the public the opportunity to raise and to have resolved in the Commission's licensing hearings important safety questions. One unfortunate consequence of the majority's decision is that it lends at least some validity to that claim. . . . The majority's attitude that the public must simply trust in the agency's ability to address the Zimmer quality assurance breakdown, particularly given the past failings of our own regulatory program in identifying these quality assurance problems, cannot help but breed public mistrust of the agency. This public mistrust is dangerous for both the NRC and industry it regulates. For, if the public lacks confidence in the fundamental fairness of our hearing process, it cannot help but question the adequacy of our regulatory program and the safety of the activities we regulate.

The majority's decision also has serious implications for the future use of *sua sponte* authority by our Licensing Boards. If the Commission will not countenance the Board's exercise of *sua sponte* authority in this case, where the issues involve serious safety matters, where those issues constitute factual disputes between the parties, and where the potential for delay is limited, it is difficult to conceive of a case in which the Commission majority *would* approve a Board's use of *sua sponte* authority. Thus, the majority's ruling may well eliminate, for all practical purposes, the *sua sponte* authority of Boards.[36]

Indian Point. A 1982 special licensing board investigation into the risk posed by the Indian Point reactors was interrupted by Commission actions unfavorable to the intervenors in the case. These reactors are just 30 miles from New York City and have a population of about 288,000 within a ten-mile radius, the area for which NRC regulations require emergency planning. Commission actions included: 1) an order requiring the licensing board to re-review contentions the board had previously

---

*The inadequacy of the alternative NRC effort at Zimmer, NRC staff investigations, is described in the section "Quality-Assurance Breakdown at Zimmer" in chapter 5.

admitted and instructing the board to reject the intervenors' testimony on the consequences of an accident unless it was combined with an analysis of the probability of such an accident[37] (the Commission subsequently refused to meet with the board to discuss the board's objections to this order[38]) and 2, reversal of the board's order allowing the silent observation of an emergency planning drill by intervenor representatives.[39] In this last order, called "silly" by Commissioner Gilinsky, the Commission found "not unreasonable" the companies' claim that the intervenor representatives (two UCS staff members) might cause overcrowding or disturbances or might pose a threat to security, despite the NRC staff's opinion to the contrary. In addition, the Commission felt that the presence of two public observers might inhibit the "spontaneity" of the exercise.[40]

According to Louis Carter, chairman of the licensing board, these Commission orders were the "final straws" that broke his "judicial back."[41] Carter recognized that requiring the litigants' testimony to include accident probability estimates dealt a severe blow to the intervenors, since such an analysis could cost from $300,000 to $1.3 million, a burden affordable only by the utilities.[42] He told the Commission that the NRC should "not blind ourselves to relevant evidence simply because the party presenting it lacks the expertise to perform a probability analysis."[43] On September 1, 1982, Carter resigned, citing as his reason the Commission's obstruction of the board at Indian Point and the Zimmer board a few months earlier. In his letter of resignation, he wrote:

> the goal of a truly independent licensing board has been needlessly subordinated to the Commission's other goals in the Indian Point case. . . . [T]o sit as the Chairman of the Board under the new restrictions and rulings would be incompatible with my sense of fairness.
>
> Unfortunately, this case has indicated to me that we do not share a common concern for the processes which regulate the resolution of these matters or in making the NRC's legal process a finer craft so that the quality of its hearings may be improved and public participation increased.[44]

In a speech to the Environmental Defense Fund Associates in October 1982, Peter Bradford, chairman of the Maine Public Utilities Commission and former NRC commissioner, recounted the Commission's interference in licensing hearings and summed up the ill effects for all parties involved:

> In a run of unfortunate decisions, the Commission majority (none of them trained in the law) reprimanded the licensing boards for raising safety issues on their own and even went so far as to reinstate a board member

who had been disqualified by the Appeal Board for making comments displaying obvious prejudice against an intervenor group. . . . The same Commission majority repeatedly declined to look into matters before hearing boards in which intervenors claimed that their rights had been infringed. Consequently, the Commission established a clear if unspoken policy of intervening only on the side of the nuclear industry. This pattern was not lost on outside observers, and I think it is fair to say that each of these three-two "victories" has cost the industry far more in terms of declining faith in the Commission's impartiality than it can possibly have gained from the actual results.[45]

The Value of Public Participation

The hearing process has long been recognized as a valuable contribution to the process of assuring the safety of nuclear plants. In 1984, the Atomic Safety and Licensing Board Panel's deputy chief administrative judge for technical matters, Frederick Shon, told the NRC's Advisory Committee on Reactor Safeguards (ACRS) that "there are some pretty high powered technical types on the intervenor side."[46] As an appeal board observed in 1974:

> Public participation in licensing proceedings not only can provide valuable assistance to the adjudicatory process, but on frequent occasions demonstrably has done so. It does no disservice to the diligence of either applicants generally or the regulatory staff to note that many of the substantial safety and environmental issues which have received the scrutiny of licensing boards and appeal boards were raised in the first instance by an intervenor.[47]

Similarly, the chief of the ASLB Panel, B. Paul Cotter, Jr., described the benefits of the hearing process in a memo to Commissioner John Ahearne in 1981:

> (1) Staff and applicant reports subject to public examination are performed with greater care; (2) preparation for public examination of issues frequently creates a new perspective and causes the parties to reexamine or rethink some or all of the questions presented; (3) the quality of staff judgments is improved by a hearing process which requires experts to state their views in writing and then permits oral examination in detail . . . and (4) staff work benefits from two decades of hearings and Board decisions on the almost limitless number of technical judgments that must be made in any given licensing application.[48]

The NRC's Rogovin report on the Three Mile Island accident also recognized the importance of the hearing process and recommended strengthening it:

Intervenors *have* made an important impact on safety in some instances—
sometimes as a catalyst in the prehearing stage of proceedings, sometimes
by forcing more thorough review of an issue or improved review procedures on a reluctant agency. More important, the promotion of *effective*
citizen participation is a necessary goal of the regulatory system, appro-
priately demanded by the public.[49]

Shon, the ASLB Panel judge, used what he termed the "policeman
analogy" to explain that even where improvements cannot be readily
identified, public participation aids the cause of safety: "You can't necessarily decide how many robberies a policeman on the beat has prevented by checking how many arrests he's made. Just his presence on
the beat discourages a lot of robberies."[50]

Public Contributions to Safety and Environmental Protection

Against the tremendous obstacles erected to meaningful public participation, intervenors have had some significant successes. Safety improvements have had some significant successes. Safety improvements
have resulted when intervenors or individual members of the public
raised issues on their own or with the aid of whistleblowers. In some
cases, improvements have been ordered by licensing boards as a result
of their verdicts on issues raised in hearings; in others, boards have issued
requirements based on agreements negotiated between the utilities and
the intervenors. Improvements have also resulted from public participation in rulemakings and other decision-making processes. Public concerns contributed to the following actions (Except where otherwise cited,
these examples were among those contained in "Reactor Safety Improvements Resulting from the Hearing Process," a list submitted by the
ASLB Panel to the Advisory Committee on Reactor Safeguards[51]):

1. Design and training improvements at St. Lucie (in Florida) to
cope with offsite power grid instabilities.

2. Improvements in the steam generator system at Prairie Island
(in Minnesota).

3. Additional requirements for turbine blade inspections and
overspeed detection at North Anna (in Virginia).

4. Improvement and conformation of the plume exposure path-
way Emergency Planning Zone at San Onofre.

5. Upgraded effluent-treatment systems at Palisades (in Michigan)
and Dresden (in Illinois).

6. A total revamping of the NRC's site-review process as a result
of contentions raised during the Pilgrim 2 proceedings (in Massachusetts).[52]

7. Safer storage for replaced steam generators at Turkey Point (in Florida).

8. Prevention of siting of the Ravenswood (New York) plant in a heavily populated area.[53]

9. Control room improvements at Kewaunee (in Wisconsin).

10. Discovery of quality-assurance breakdowns at Zimmer, Midland (in Michigan), and South Texas.

11. Upgraded requirements at Beaver Valley (in Pennsylvania) for steam generator tube-leakage plugging.

12. Expanded dose monitoring requirements at Davis-Besse (in Ohio).

13. Detailed upgrading and additional requirements at Shoreham (in New York) affecting many plant systems and procedures.*

Public contributions in generic matters include:

1. Improvements in the specificity of the requirements for the evaluation of emergency core cooling systems.

2. New guidelines on off-site radioactive exposures to be kept "as low as practicable" or approximately 1 percent of original limits.

3. Reanalysis of steam and high-pressure line routing to reduce the dangers of pipe rupture outside the containment and damaging of safety systems.

4. Closer examination of guidelines for determining distance and activity of earthquake faultlines on acceptability of location of reactors.

5. New regulations to improve the protection of safety equipment against fires.[54]

6. New regulations to ensure that plant safety equipment can function in a harsh accident environment.[55]

7. Improvements in NRC guidelines and operating practices of licensees and contractors in the areas of quality control and quality assurance.

8. Uncovering weaknesses in plant security requirements.

9. Closer attention to the impact of fuel-pellet densification on the safe operational level of certain boiling water reactors.

---

*Systems and procedures affected include the remote-shutdown panel, the diesel-generator relays, the detection of inadequate core cooling, the loose-parts monitor, emergency core cooling pump blockage, fire protection, human factors engineering, containment isolation, preservice inspection, inservice inspection, as low as reasonably achievable (ALARA) practices for maintenance procedures, post-accident monitoring, core spray pump operation, cable separation, surveillance testing of cable penetrations, safety system annunciation, turbine-rotor surveillance, fuel-pool leakage, and reactor cooling system leaking.

10. Greater use of closed-cycle cooling towers and ponds to lessen heated discharges into rivers and lakes.

11. Increased attention to the problems of fish entrapment and marine life entrainment caused by design and location of a plant's cooling water intake.

12. More careful review of the effects of the release of radioactive materials on marine life, shellfish, and clam beds.

13. Fundamentally improved approach to environmental assessment because of the *Calvert Cliffs* case.

The Cost of Ignoring Intervenors

Although a construction permit is granted on the basis of only preliminary design information, the level of NRC technical review drops dramatically once the permit is issued. The on-site inspector, occasionally augmented by other NRC personnel, can audit only a tiny fraction of the construction programs, and the audit consists largely of reviewing paperwork. Only about 1 percent of the as-built plant is physically inspected before operation. As a result, quality-assurance breakdowns have come to light through the revelations of plant workers and have been pursued by intervenors.

Unfortunately, there have also been many cases where the public's concerns were ignored, only to resurface later as actual problems. By that time, after construction was far along or operation had begun, fixing the problems was inevitably more difficult; in some cases, it was economically infeasible. Five cases provide prominent examples.

Zimmer. A long list of quality-assurance problems at Zimmer was brought to the NRC's attention three times between 1976 and 1980, with little result. The extent of problems discovered in 1981–1982 finally led the Commission to vote in November 1982 to stop all work at the plant. At that point the plant was considered to be 97 percent complete, $1.7 billion had been spent, and the owning utilities could not afford the necessary work to fully inspect the plant and determine its condition. The plant was canceled instead, and conversion to coal was being considered, although the cost would be enormous. The financial stranglehold the utilities eventually found themselves in could likely have been avoided if the NRC had not for years closed its ears to evidence of problems at the plant.*

---

*See "Quality-Assurance Breakdown at Zimmer" in chapter 5 for further details and references.

Diablo Canyon. Between 1977 and 1980, intervenors unsuccessfully tried more than once to raise the issues of design and construction quality assurance at Diablo Canyon. The NRC eventually had to suspend Diablo Canyon's low-power license two months after it was issued, when the disclosure by the utility of an error in seismic design led to investigations that further disclosed a widespread breakdown in the quality-assurance program.[56] The subsequent reverification program resulted in thousands of modifications to the design of the piping and supports.[57]

TMI-2. Intervenors challenged the viability of the official emergency plans at TMI-2, but the licensing board found them to be both "adequate and workable."[58] When the plans were called upon less than eighteen months later, during the March 1979 accident, they proved to be non-existent or inapplicable.[59]

Seabrook. During and after construction permit hearings for Seabrook, intervenors repeatedly questioned the financial ability of the principal owner, Public Service Company of New Hampshire (PSNH) to complete the project.[60] Their concerns were ruled invalid. In 1984, PSNH, facing bankruptcy, was forced to stop all work at Unit 2 and turn over the fate of Unit 1 to a new, highly uncertain financing entity.

Shoreham. During the Shoreham construction permit hearings, intervenors on Long Island raised the issue of the difficulty of evacuating the area of the proposed plant site. The licensing board ruled in 1973 that a detailed emergency plan need not be prepared or considered until the plant was completed and ready to operate. The intervenors accurately noted that this "foreclosed consideration of whether *any* effective emergency plan could be developed for Long Island. . . . "[61] A later confrontation—centered on Suffolk County's conclusion that the plant should never operate because the area could not be successfully evacuated—and the prospect of having spent over $4 billion in vain might have been avoided if the utility and the Atomic Energy Commission had taken the issue seriously in 1970.

A Reactor with No Intervenors

The case of the Grand Gulf 1 reactor, where there were no intervenors, illustrates the effect that a lack of public scrutiny may have. Representative Edward Markey, chairman of a House Interior and Insular Affairs subcommittee responsible for overseeing the NRC, said that after the NRC granted the plant an operating license, the licensee, Mississippi Power & Light (MP&L), and the NRC staff discovered that the conditions specified in the reactor's license contained more than a thousand errors. They also found that operators employed at the plant were unqualified, training records were apparently falsified, and there were design errors in a number of safety systems.[62] In addition, the staff apparently disregarded the Atomic Energy Act and NRC regulations by allowing the reactor to operate for a time without an adequate source of emergency electrical power and failing to issue a license amendment that would have required public notice and opportunity for a hearing. At the same time, however, the Commission *did not* allow the Shoreham plant, opposed by intervenors with unusual technical and financial resources, to operate without the same equipment.*

It is fair to argue that what Grand Gulf needed was a healthy dose of intervention. Markey concluded, "I cannot help but think that such laxness would not have been possible if there had been intervenors and licensing boards reviewing the work of the NRC staff and MP&L and that these errors would have likely been detected prior to licensing."[63]

NRC Efforts to Curtail Public Participation in the Licensing Process

Even though public scrutiny has enhanced plant safety, the NRC has made various attempts over the years to "expedite" or "reform" the licensing process by restricting public participation. These efforts involved raising the specter of licensing delays where none existed, proposing legislative changes, and proposing changes in administrative procedure.

"The Great Licensing Delay Hoax of 1981"

Less than a year after the Kemeny commission criticized the NRC's focus on licensing to the neglect of safety problems and only months after the NRC's own Rogivin report labeled the hearing process a "sham" and recommended steps to strengthen it, the NRC was swept up in a campaign to bypass the hearing process because of predicted licensing delays. In the end, the plants that the NRC and the industry predicted

---

*See chapter 5, section on "Grand Gulf: Misuse of the Standardized Design Concept and the Perils of Self-Regulation," and chapter 4, sections on "Double Standards at Grand Gulf" and "Ignoring an Old and Inventing a New Regulation at Shoreham."

would be delayed were not delayed. In what Commissioner Bradford called the Great Licensing Delay Hoax of 1981,[64] safety issues again took a back seat to the goal of speedy licensing.

It began with an approximately thirteen-month pause in licensing taken by the NRC after the TMI accident in order to determine what changes were needed for new plants. In April 1980, soon after the NRC resumed granting licenses, Chairman Ahearne told a House Appropriations subcommittee that the construction of three plants would be completed before the NRC's licensing process would be concluded, and a total of eleven months of operating time would be lost.[65]

Nuclear industry lobbyists—encouraged and aided by Representatives Tom Bevill and John Myers, chairman and ranking minority member of the House Appropriations subcommittee that determines NRC's budget—used Ahearne's testimony to mount an extensive public relations campaign around the threat of looming "licensing delays."* By January 1981, the projected delays had risen to 79 months for eleven plants[66] and an anonymous industry source was quoted in a trade journal as estimating that these delays would cost ratepayers more than $4 billion.[67]

Bowing to pressure caused by these alleged delays, the NRC commissioners by the end of February began to meet almost daily to discuss ways to "expedite" the licensing and hearing process. In March, the Commission sent Congress a "legislative reform" package that included a draft bill granting a two-year authorization for the NRC to issue "interim" licenses—*before* hearings were concluded and all safety issues resolved—for operation at up to 5 percent of full power.

William Lee, chief operating officer of Duke Power Company, testifying on behalf of three industry groups, told the Senate Subcommittee on Nuclear Regulation chaired by Senator Alan Simpson that thirteen plants were likely to be delayed a total of 90 months. This "regulatory delay" was to cost between $2.7 and $3.6 billion. In addition, Lee warned: "If this is extrapolated to all the plants expected to be delayed through

---

*In late 1980, at Bevill's direction, the NRC began to submit to the subcommittee monthly reports totaling the estimated months of "delay" between estimated reactor construction completion dates and estimated NRC approval dates for operation. In February 1981, Bevill directed the NRC to use the utilities' construction completion date estimates, rather than the NRC staff's. The effect of this change on estimated delay times was illustrated by the March 1981 report that included two estimated delay figures: 71 months by the NRC and 110 months by the industry. In the following month's report, the NRC for unknown reasons dropped the staff's estimates altogether; only the utilities' estimates were used from then on. These reports, which came to be known as the Bevill reports, are still being issued, but on a quarterly basis.

1983, the total cost of delay would be \$7 to \$10 billion."[68] He concluded that the NRC's interim licensing proposal would not cure the problem and supported NRC Chairman Joseph Hendrie's position that such a license should allow a plant to proceed to full power. The Union of Concerned Scientists told the same subcommittee that these delay figures were based on estimated construction completion dates, which in the past had proved to be optimistic, and that there was no evidence that the hearing process caused any delays.[69] Nevertheless, Congress passed an amendment allowing the NRC to grant full-power "temporary operating licenses." In the months that followed, UCS's prediction proved correct. Construction completion dates slipped further into the future. In addition, the NRC adopted some administrative measures to move the licensing process along. The delays completely evaporated, and the interim licensing authority passed by Congress to meet the supposed emergency was never used. For example, three of Lee's estimated construction completion dates were March 1981 for Diablo Canyon Unit 1, September 1982 for Shoreham, and November 1981 for Zimmer. According to the NRC, Diablo Canyon was "completed" in April 1984, Shoreham was "completed" (for low-power operation) in May 1984, and Zimmer was canceled in January 1984.

Besides pressing Congress for the interim operating licensing authority (even as the delay figures were clearly revealed to be unfounded), the NRC took some internal actions that undermined safety. In March 1981, the House Subcommittee on Energy, Environment and Natural Resources, chaired by Representative Toby Moffett, undertook an investigation of the impact on nuclear safety of the NRC's licensing speed-up proposals. As a result of this investigation, the House Government Operations Committee issued a report in October 1981 saying that while the NRC was devoting personnel and resources to "the alleged problem of licensing delay," its safety-related projects had slipped behind schedule. One major set of these projects was the TMI Action Plan, the changes to correct generic safety problems shown to be necessary by the TMI accident. A 1981 internal audit of the NRC had found that over the previous year the Commission had not effectively tracked the implementation of the post-TMI requirements. Now the House committee found that other safety-related programs—the program to evaluate the safety needs of the oldest plants, the fire-protection requirements, the equipment qualification program, the unresolved safety issues program—had also fallen behind schedule. The committee report concluded:

The NRC response to the alleged problem of licensing delay has been misguided. Its action and proposals may be detrimental to safety over the long term, and they risk undermining public confidence in the Commission and the licensing process.

Safety-related NRC programs and the inspection of operating nuclear reactors—the true mission of the NRC—have remained behind schedule at the same time the NRC's attention has been devoted to the supposed problem of licensing delay.[70]

Licensing Delay Scares of 1984

Unfortunately, the NRC soon repeated its mistake. On March 16, 1984, Chairman Nunzio Palladino held an unannounced meeting with Commission-level personnel (executives of the Office of General Counsel and the Office of Policy Evaluation) and the highest-level NRC staff (the executive director for operations and the executive legal director). The chief administrative judge of the NRC's licensing boards, B. Paul Cotter, Jr., was also present. At this meeting, ideas for "expediting" a low-power licensing proceeding for the Shoreham plant on Long Island were discussed. (As Commissioner Gilinsky later noted, Palladino was urging that ways be explored to authorize low-power operation at a time when the utility had not yet submitted a request for low-power operation.[71]) Palladino held this meeting even though the Administrative Procedure Act and the Commission's regulations prohibit the commissioners—the ultimate judges in licensing proceedings—from meeting with any one party to a proceeding without granting all the other parties the opportunity to be present.[72] The NRC staff, of course, was a party to the Shoreham proceeding. Palladino did not invite the other commissioners nor even inform them of the meeting.

Four days later, on March 20, Palladino warned the other commissioners of possible delays at four plants and "difficulties" being experienced at five others. He asked for a special meeting to discuss these problems and consider options to "expedite" the respective licensing proceedings.[73]

That same day, Long Island Lighting Company(LILCO), the Shoreham operator, requested that a low-power license be issued on an expedited basis. Two days later, before the other parties had an opportunity to respond to LILCO's motion, Palladino's legal assistant called Cotter—the ASLB judge who had attended Palladino's March 16 meeting— and read him a "working paper" indicating that the Commission wanted the low-power issue decided within 30 days. The next day, Cotter provided his comments on the working paper in the form of a draft licensing board order expediting the proceeding under a schedule Cotter reportedly termed "brutally tight" and "definitely not recommended."[74] Palladino later claimed that Cotter's proposed expedited schedule was done on Cotter's own initiative and was not influenced by Palladino.[75]

On March 30, Cotter appointed a new licensing board to hear LILCO's request for a low-power license. Schedule conflicts were given as the reason for appointing a new board.[76] However, the old board had rejected similar arguments for allowing low-power operation without demonstrating the reliability of on-site emergency power. It was not until April 4 that Palladino provided the other commissioners with his working paper and the draft order. Also on April 4, the intervenors told the licensing board that the expedited schedule would severely prejudice their ability to present their case.

Two days later, the new board issued a schedule that allowed only eighteen days between the start of discovery and the start of the hearing, including three weekends, Passover, Palm Sunday, Good Friday, and Easter. The Schedule was notably similar to that laid out in the draft order written by Cotter in response to Palladino's working paper.[77] The board offered no rationale for the tight schedule, lending the appearance that it had adopted the rationale urged by LILCO: "The Shoreham Nuclear Power Station represents both a huge commitment of economic resources and Long Island's only power plant not dependent on foreign oil. Thus, there are compelling reasons for the station's early operation."*[78]

In the next several days, Palladino came under congressional fire. Representative Markey wrote a series of letters to Palladino expressing concern over the "appearance of impropriety" in Palladino's actions regarding the Shoreham case. Markey suggested that Palladino, having given the appearance of prejudging the Shoreham case, should consider disqualifying himself from further decision making in the case.[79] Palladino responded that he was only fulfilling his responsibility to ensure that licensing proceedings would be carried out "within a reasonable time" as required by the Administrative Procedure Act.[80]

Palladino's fellow commissioners also were critical. On April 24, at the meeting Palladino had requested the previous month to discuss ways

---

*See chapter 2, section on "Cost-Benefit Analysis," for a discussion of legal limitations on the NRC's authority to consider costs in making safety decisions.

to "expedite" licensing proceedings, Commissioner Gilinsky strongly objected to the chairman's attempt to repeat the Commission's past mistakes: "The fact of the matter is that in the past these delays have evaporated . . . the agency has panicked at the sight of these numbers, delay estimates, and has I think behaved in a way that it shouldn't have, changed its procedures and resource allocations I think to the detriment of overall safety regulation."[81]

Commissioner Asselstine said that these newly alleged delays were based on the utilities' construction completion dates, which had historically been unrealistic. He also said that it was incorrect to call them delays:

> [Nobody] is in favor of unnecessarily delaying the operation of these plants but in my view an unnecessary delay is not a delay caused by the need to resolve a significant safety question whether that is by the staff or in a hearing.
>
> To characterize some of these cases as delay cases . . . is inaccurate because the extended proceedings have in some measure been caused by the need to resolve significant safety issues either by the staff or in hearings or both.[82]

On April 25, a U.S. District Court disagreed with the view that the NRC's hearing schedule in the Shoreham case was reasonable and issued a temporary restraining order prohibiting the NRC from going ahead with the schedule. The court found that "meaningful participation in the administrative proceeding by plaintiffs has been precluded by the limited discovery period."[83] Responding to the court order, the Commission vacated the licensing board's schedule on April 30 and established a new one on May 16, this time allowing 53 days between the start of discovery and the hearing.

Representative Morris Udall, chairman of the House Interior and Insular Affairs Committee, called a subcommittee hearing on May 17 in which he said: "We are . . . very much concerned with the integrity of the regulatory process. In this case it appears that procedures adopted to resolve substantive safety issues did not provide adequate protection of the rights of all parties."[84]

Commissioner Gilinsky highlighted the impropriety of the chairman's private meeting with the staff:

> The Staff is a party in the hearing; the Chairman is one of the ultimate judges. The Staff Directors should have told the Chairman politely that it is not their job to carry the ball for the Company. It is understandable that they did not say this under the circumstances. The Chairman is, by law, the Staff's direct supervisor. He controls annual bonuses worth many thousands of dollars to senior Staff members. What we have is a situation

in which one member of the ultimate NRC adjudicatory tribunal appears to be directing the actions of a key party in the case.[85]

Commissioner Asselstine pointed to "procedural irregularities associated with certain actions by the Chairman" and the conduct of the new licensing board chairman, which, "taken together . . . create the appearance of impropriety in the conduct of this proceeding." While he was pleased that the Commission had remedied the board's order, he regretted that the Commission had not addressed the procedural questions: "The consequences of the majority's inaction are enormous and far-reaching. By its inaction, the majority undermines the credibility of our licensing hearings and the integrity of our entire regulatory program."[86]

In June 1984, Suffolk County and New York State formally requested that Palladino disqualify himself from the Shoreham proceeding.[87] Palladino announced that he would do so temporarily and requested the NRC staff's opinion before making a final decision. The staff argued against disqualification. In September, Palladino maintained that he had done nothing improper and announced that he would continue to vote in the Shoreham case.[88]

Thus, this episode had precisely the opposite effect of that intended by the chairman. Rather than "expediting" the Shoreham license, the proceeding was halted temporarily by court order and a new set of questions concerning the integrity of the NRC came to the forefront.

In the long run, however, other questions were to determine whether Shoreham went into commercial operation. Suffolk County concluded, through extensive study, that safe evacuation from the areas surrounding Shoreham was impossible and refused to participate in emergency preparedness plans, which it viewed as a sham. New York State supported the county's decision. Without their participation, Shoreham faced difficulty getting a full-power license.

In April 1983, the licensing board stated that in spite of NRC regulations that allow low-power operation before an off-site emergency plan is approved, "as a matter of sound public policy" low-power operation should not be permitted "in circumstances where there is no reasonable assurance" that the plant will ever be granted a full-power license.[89] The board asked the Commission to review this question. In June, the Commission (with Asselstine and Gilinsky dissenting) ruled that Shoreham could be granted a low-power license in spite of substantial doubts about whether it could ever receive a full-power license. According to Gilinsky, the Commission made this decision without reviewing the details of the

Shoreham situation to conclude whether the fears over the uncertainty
of the full-power license expressed by the licensing board were exag-
gerated. The Commission also made this decision without allowing the
parties to comment. Gilinsky charged the Commission with "playing
chicken with the Governor of New York:"

> The Commission sought to justify its actions with the familiar pretense
> that the utility proceeds "at its own risk." In actuality, the utility proceeds
> at the risk of the public. In light of the fundamental change in plant
> conditions which results from the irradiation of fuel, and of the associated
> substantial increases in the cost of maintenance and plant modification,
> the common sense and responsible view is that a low power license should
> not issue when there is substantial question that full power operation of
> the reactor will ever be permitted.[90]

LILCO faced bankruptcy because financial institutions were unwilling to extend more credit for construction of Shoreham. The company needed permission to operate at low power in order to send a
positive message to the investment community.[91] The attempts by Palladino and the licensing board to expedite a low-power license for Shoreham made it appear that the disposition of the investment community
weighed heavily on their minds as well.

As in 1981, the 1984 panic over licensing delays caused a diversion
of NRC resources from important safety responsibilities to the issuing
of reactor operating licenses. In May 1984, Gilinsky sent a memorandum
to Palladino expressing his concern that the safety of operating plants
in the NRC's Region IV had been affected by reassignment of inspectors
and support staff to work on getting Waterford and Comanche Peak
licensed.[92] Asselstine expressed even broader worries:

> I believe that the concern about drawing resources away from operating
> plants applies to all of the Regions, and not just to Region IV. In fact, in
> my recent visits to two regions, regional staff has been quite candid about
> this concern, and have agreed that the emphasis on [new licenses] is having
> a significant effect on our ability to assure safety at the operating
> plants.[93]

At a meeting to discuss these concerns, Region III Administrator
James Keppler said that although he felt the situation was "acceptable,"
the regional inspection enforcement staff was "about as stretched as . . . I
want to be right now."[94] In addition, the NRC's director of licensing,
Darrell Eisenhut, revealed that "so far this year we are about nearly 20
percent over in terms of our expenditures for operating license reviews
over what we projected" and 13 percent under projected expenditures

for operating reactors, including "following up on occurrences [malfunctions and near-accidents]."[95]

William Dircks, head of the NRC staff, was asked by Palladino to review the situation. He told the chairman that he shared the concern that reallocation of resources "has the potential for adversely impacting our operating reactor program responsibilities." While he believed that attention to operating plants had not been reduced to "an unacceptable level," there had been "a general increase over the past year in Staff resources to [new operating license] reviews (and inspection effort), and a related decrease in our allocation of effort to operating reactors."[96]

Legislative Proposals

The interim licensing authority requested of Congress in 1981 was only one of a host of proposals that the NRC has made to curtail public participation in the licensing process. These proposals for changes in the licensing and public hearing process—licensing "reform"—would further restrict public access.

For example, the NRC's proposal sent to Congress in 1983 would have eliminated the right to cross-examine witnesses under oath, a right now protected by the Administrative Procedure Act "as may be required for a full and true disclosure of the facts."[97] Instead, questioning of the utility and staff witnesses would be allowed only with the permission of the licensing board if the intervenors first could prove that there were "genuine and substantial disputes of fact which can be resolved with sufficient accuracy only by introduction of evidence at a formal hearing" and that the Commission's decision was "likely to depend in whole or in part on the resolution of the dispute."[98] However, since the intervenors would have no right to cross-examine the utility's witnesses or the NRC staff before making this case, a licensing board would in effect be making a decision on the issues without hearing questioning of the utility or the staff.

Another proposal would also allow the NRC to issue a combined construction permit and operating license without a final design in hand. In essence, the NRC would be granting an operating license before construction begins, while knowing no more details about the plant design than it currently knows before granting a construction permit—preliminary, conceptual information. Gilinsky likened this type of "one-step" license to "handing an incoming freshman his college diploma on the basis of his course outline."[99]

The NRC's legislative proposals have contained a provision for the approval of "standardized designs"—designs that, once approved, could be used by applicants for construction permits without the need for a new NRC approval or hearings. If final designs—with no unresolved generic safety issues—were required prior to approval, standardization could enhance safety and public participation by allowing for scrutiny of the design before plants have been largely built. In fact, the NRC has long had the legal authority to approve standardized designs and has rules in place to do so; no more legislation is needed. But the NRC's handling of one such standardized design that was up for approval, the General Electric Standard Safety Analysis (GESSAR), raised serious questions as to whether safety would be enhanced by its use. The design proposed in GESSAR did not resolve all the generic problems applicable to boiling water reactors. Furthermore, General Electric performed a probabilistic risk assessment (PRA) that claimed the risk from the reactor was negligible, and GE used this PRA to argue against design changes recommended by some NRC staff members. In response to a Freedom of Information Act request filed by UCS, GE argued that the key portions of the PRA were "proprietary." The NRC supported this claim, refusing to allow public release of hundreds of pages of technical analyses relied upon by GE. Subsequent to the filing of a Freedom of Information Act lawsuit by UCS, much of this material was released.

The NRC has devoted hundreds of hours of staff time to developing licensing "reform" proposals. In the meantime, serious safety issues have been allowed to linger unresolved, quality-assurance disasters have caused largely completed plants to go under, and precursors to severe accidents have occurred in large numbers. According to former Commissioner Bradford, at the very moment that Secretary of Energy James Schlesinger was testifying before a House subcommittee urging "expedited licensing," NRC Chairman Joseph Hendrie was advising the governor of Pennsylvania to evacuate pregnant women and young children from within five miles of Three Mile Island.[100] The TMI–2 accident cut short efforts to pass that year's licensing reform bill.

Bradford, testifying in July 1983 on behalf of the National Association of Regulatory Utility Commissioners, told a Senate subcommittee that "we are under no illusion that the problems of nuclear power can be solved by [this kind of] licensing reform legislation." Instead, Bradford recommended that licensing reform legislation include a provision for "[a] safety issue resolution process at least as firm as the NRC's commitment to meeting licensing deadlines." He observed:

NRC hearings did not cause Three Mile Island. NRC hearings did not bring about the cancellation and default at the WPPSS units. NRC hearings had nothing to do with the quality assurances breakdowns at Diablo Canyon and Zimmer. NRC hearings are not causing the Midland containment to sink. NRC hearings are not even at the bottom of the cost overruns st Shoreham and Seabrook.

Most of nuclear energy's real problem situations would tend to support the proposition that, if anything, the NRC process of nuclear surveillance has been too lax rather than too stringent. The problem that Congress should address first is not that the United States has licensed too few plants too slowly; it is that we have licensed and made large financial commitments to too many plants too quickly. That is why we now have a landscape dotted with 9 figure cost overruns, a 9 figure accident, 8 figure cancellations and 8 figure mishaps in such areas as steam generator tubes, pressure vessels, emergency plans, scram systems, and quality assurance.[101]

Administrative Proposals

The most severe set of changes to the hearing process thus far proposed publicly by the NRC was that written by the Regulatory Reform Task Force, an NRC staff committee chaired by James Tourtelotte. The NRC assumed that these administrative changes, proposed for public comment in 1984, could be achieved without new statutory authority. In UCS's view, however, they were inconsistent with both the Atomic Energy Act and the Administrative Procedure Act.

One provision would require intervenors to submit, along with their initial petition to intervene, their contentions and an identification of the documents and expert opinions they would use during the hearing. In essence, that means they would be required to present their entire evidentiary case before even being admitted to a proceeding and before the NRC released its own safety evaluation of the plant—and all within 30 days after receiving notice of a hearing. Gathering evidence to support a case, particularly on complex technical issues, generally takes several months, depending on the number of issues. The new NRC proposal would require intervenors to digest thousands of pages of technical data, identify all significant issues, seek out and obtain experts, and plan expert testimony within the 30-day limit—a nearly impossible task. The NRC staff, with its technical and legal resources, routinely takes many months to review the technical data and prepare its safety evaluation report.

The right of intervenors to cross-examine would also be tightly limited by the task force proposal. Intervenors would have to file a special request, specifying why the points could not be made by direct testimony as opposed to cross-examination, and submit detailed cross-examination plans within ten days after receiving written testimony. While utilities and the NRC staff have the resources to introduce hundreds of pages of direct testimony, mosts intervenors must rely on cross-examination to reveal the basis underlying this testimony, which otherwise might appear incontrovertible. In addition, the proposal would limit cross-examination (where granted) to issues of fact, thereby eliminating questions related to credibility, bias, opinion, and the qualification of experts—areas where questioning is often crucial.

The task force also suggested a sweeping provision that would allow the Commission to "prescribe such [hearing] procedures as it deems necessary." Currently, the Commission is allowed to dictate some hearing procedures but must change procedures specified by rules through a rulemaking process that requires soliciting and considering public comment before altering rules. The process that governs the actions of all federal agencies is established by the Administrative Procedure Act and would be lost under the task force proposal, which would allow the Commission to change the fundamental procedural rules of the game at its will.

The commissioners apparently were surprised by the degree of dissent the proposal aroused among some task force members, the chairman of the appeal board panel, and other senior NRC personnel. It was suggested that the proposal could be sent out for public comment without an endorsement. Peter Crane, of the Office of General Counsel, a task force member who did not endorse the package, said that such a move would put a "blessing" on the proposal and "could give members of the public a serious question about the Commission's judgment and the Commission's approach to its regulatory responsibilities." Commissioner Gilinsky said that "this whole package has a sort of air or character of bashing intervenors and constraining participants." He agreed that putting it out for comment would indicate that it was what the Commission had on its mind and might very well carry out.[102] The proposal, described as containing "suggestions for improvements in the licensing process" but "not Commission proposals," was published, nevertheless; and, while the notice acknowledged cryptically that there was disagreement within the task force, the description of the proposal gave no hint of the potential disadvantages of the proposed changes. There was no acknowledgment, except in Commissioner Asselstine's separate views, that the effect of this licensing "reform" would "as a practical matter effectively eliminate public participation in the hearing process."[103]

The NRC's Response to the Government in the Sunshine Act

A measure of the NRC's resistance to public scrutiny is its attitude toward the law that requires it to conduct its business in the open. The Government in the Sunshine Act, in effect since early 1977, regulates meetings held by government agencies with two or more presidential appointees. The act requires all meetings to be open to the public unless the subject matter is specifically exempted under one of the act's ten exemptions, thus ending the authority of boards, commissions, and authorities to meet and make decisions in private.

Congress framed the act on the theory that "the greater the openness in government, the greater the public trust and confidence in government." Other associated benefits envisioned were stimulation of broader participation in agency processes, more extensive debate of agency policies, agency responsiveness to a broader array of interests, stronger accountability, reduced public misunderstanding resulting from partial information, improved rates of compliance with agency regulations, and higher quality of work resulting from greater public scrutiny.[104]

A study of the implementation and effects of the Sunshine Act reported some interesting findings regarding the NRC's attitude. The study, released in June 1984, surveyed government agency members and upper-echelon staff from eighteen major agencies and reported their responses regarding two issues that indicated attitudes toward the act: the act's effect on the agencies' ability to work with Congress and other agencies, and its effect on the agencies' decision-making processes. Of the eighteen agencies, the NRC displayed the second most negative attitude toward the act in its response on the first issue (the Securities and Exchange Commission had the worst) and the most negative attitude of all in its response on the second.[105]

The study also surveyed the agencies' "attentive publics"—journalists, attorneys, and trade, labor, and public-interest groups that deal with the agencies. For most agencies, their publics' perceptions of the effects of the Sunshine Act were not far apart from the agencies' own perceptions. The most striking discrepancy between the two groups existed for the NRC: "its officials see a distinct impairment, whereas its public reports

that effectiveness is enhanced by the act."[106] The study attributed the
agency's discomfort with the act to "the dispositions of agency leadership"
and "age and tradition." In particular, "the NRC is a much older agency
whose functions and organizational history going back to the Atomic
Energy Commission create a special concern for the confidentiality of
information."[107]

The study identified some specific problems in the NRC's implementation of the Sunshine Act. While most agencies typically met the
seven-day requirement for placing advance notice of meetings in the
Federal Register, the NRC failed to meet this standard approximately
half the time during the 1977–81 period.[108] The Commission has also
on a number of occasions fought to close meetings when closure was
either illegal or not in the public interest. On at least two occasions, the
NRC has been found by the courts to have violated the Sunshine Act.
In one case, Common Cause, a national consumer-advocate organization,
sued the NRC for closing its budget deliberations.[109] In the second case,
the *Philadelphia Inquirer* sued the NRC for closing its deliberations on
the restart of TMI-1.[110]

In May 1985, without prior public notice or opportunity for comment, the NRC published an immediately effective interim rule to exclude the press and public from more of its meetings. The rule created
a broad new class of nonmeetings at which, although official agency
business is discussed, the talk is not expected to be "sufficiently focused"
to cause individual commissioners to "form reasonably firm positions."
Beyond the question of how one can determine in advance the likelihood
that some commissioner may form a "reasonably firm position" during
the course of a meeting, the most troubling aspect of the rule is that,
unlike current practice, no transcripts of the secret sessions are kept.
Neither the public nor reviewing courts have any way of knowing what
was said. Should a court later rule that the NRC illegally closed a meeting,
there would be no record of the discussion available for public release.
Thus there will be no effective legal recourse against mistakes or abuses
of the rule.

FOUR

ARBITRARY ENFORCEMENT
OF REGULATIONS

> The judgment of the Commission . . . was
> that public health and safety was adequately
> protected. That's where we start here. And
> then we go to parsing and picking at regula-
> tions.
>
> Commissioner Frederick Bernthal

The NRC's treatment of its regulations and the law have suffered when the agency's *de facto* priorities—licensing reactors, keeping them on line, and minimizing financial hardships on utilities—are at stake. NRC regulations, when they obstruct these objectives, tend to disappear suddenly like the Cheshire Cat in *Alice in Wonderland*, leaving only the Commission's grin. When a problem could not be put aside by labeling it generic (because the issue has already been formally resolved in a regulation) or a hearing might have impeded speedy licensing, the NRC has sometimes simply ignored its regulations. In other cases, where plants do not meet the regulations, the NRC has hastily changed its regulations to match the plants. Other NRC actions have shown disrespect for the legal requirements of the federal Administrative Procedure Act and the Atomic Energy Act.

In seeking to avoid requiring a utility to comply with a regulation, the NRC has argued, explicitly or implicitly, that the plant is safe even though the regulation is violated. That is a continuation of a practice used by the AEC. However, NRC regulations represent the Commission's definition of the minimum standards required to protect the health and safety of the public. Thus, full compliance with the regulations is a necessary and sufficient prerequisite to issuance of a license to construct or operate a nuclear power plant. This principle was articulated by an AEC appeal board in rebuffing the staff's attempt to allow the Vermont Yan-

kee plant to operate despite the fact that its emergency core cooling
systems did not meet regulatory requirements:

> an intervenor cannot validly argue on safety grounds that a reactor which
> meets applicable standards should not be licensed. By the same token,
> neither the applicant nor the staff should be permitted to challenge ap-
> plicable regulations, either directly or indirectly. Thus, those parties
> should not generally be permitted to seek or justify the licensing of a
> reactor which does not comply with applicable standards. Nor can they
> avoid compliance by arguing that, although an applicable regulation is
> not met, the public health and safety will still be protected. For, *once a
> regulation is adopted, the standards it embodies represent the Commission's definition
> of what is required to protect the public health and safety.*
>
> In short, in order for a facility to be licensed to operate, the applicant
> must establish that the facility complies with all applicable regulations. If
> the facility does not comply, or if there has been no showing that it does
> comply, it may not be licensed.[1]

The NRC has an exemption process through which utilities can
demonstrate that compliance with a regulation is truly unnecessary. Yet
the agency has been willing to sidestep this process as well.

A vivid illustration of the NRC's attitude toward its regulations and
the law is the equipment qualification story.* When it became clear that
utilities would be unable to meet the deadline on the rule requiring
qualification of safety equipment, the Commission simply waived the
deadline. In other words, it changed the rule to meet the needs of the
utilities rather than forcing the utilities to change their plants to meet
the regulations. The NRC also flouted its regulations and the law by not
allowing an opportunity for public comment on this action. The appeals
court's ruling that the NRC's action in this case violated the Atomic
Energy Act, the Administrative Procedure Act, and its own rules was
one of five court decisions against the NRC in a fourteen-month period.[2]

This chapter discusses other examples of trampled regulations and
clashes with the law that have occurred in the areas of emergency planning, fracture toughness of reactor vessels, reactor operator qualifications, emergency power supplies, the exemption process, the hearing
process, and the license amendment process. The fact that the NRC does
not actually know which plants meet which regulations is also discussed.

Indian Point: A Mockery of the Rules

A lengthy saga at the Indian Point plant near New York City demonstrated that the Commission is unwilling to enforce its own regulations.

---

*See chapter 2, section on "Equipment Qualification."

In this case, the agency failed to uphold its emergency planning regulation at the most populous site in the United States.

In August 1980, almost seventeen months after the emergency plans at TMI-2 were shown to be inadequate when called upon, the NRC published new, strengthened off-site emergency planning regulations requiring all operating reactors to comply by April 1, 1981. The rule stated that if the emergency plans for a licensed reactor were found to be deficient by the Federal Emergency Management Agency (FEMA, the agency responsible for reviewing them) the utility would be given a 120-day grace period to bring the plans into compliance. If the deficiencies were not corrected after this period, the Commission would determine whether reactor shutdown or other enforcement action would be appropriate. This determination was to be based on whether the utility could demonstrate that the deficiencies were "not significant," that "adequate interim compensating actions have been or will be taken promptly, or that there are other compelling reasons for continued operation."[3]

When the April 1981 deadline arrived, FEMA found numerous "deficiencies" in the plans for the Indian Point nuclear reactors,[4] 30 miles north of New York City. The 120-day clock was started for Consolidated Edison (Con Ed) and the Power Authority of the State of New York (PASNY), the owners of units 2 and 3 respectively.*

In August, as time was running out, the utilities gave revised plans to FEMA. In an interim report requested by the NRC, FEMA wrote that planning was generally adequate but that a final determination could not be made until exercises were held.[5] The following week, the NRC declared that deficiencies at Indian Point had been resolved satisfactorily.[6]

In March 1982, an exercise of the plans was held. More than four months later, FEMA concluded that the exercise had revealed "significant deficiencies."[7] Instead of shutting the plant down, the Commission, on August 3, started *another* 120-day clock. When time again ran out in December, FEMA issued a report that said Indian Point's plans were

---

*In September 1979, UCS had petitioned the NRC to shut down the Indian Point reactors unless and until the NRC could determine after hearings that the potential consequences of a serious accident would not be so severe as to make the site unsuitable and that a number of reactor safety improvements were made. Sixteen months later—after more than 100 comments were sent to the Commission in support of UCS's petition, along with letters of support from civic and political leaders, and after extended deliberation on what the scope of hearings should be—the Commission ordered hearings. The hearings were to determine the status of Indian Point's emergency plans, the risk posed by the reactors particularly in comparison with other factors, and the environmental and economic consequences of an Indian Point shutdown. The reactors were allowed to remain in operation while the hearings were in progress.

still inadequate in a number of respects.[8] However, FEMA declined to assess whether the plans could be implemented until after another drill could be held the following March. Meanwhile, in May 1982, Rockland County determined that the plans in place, which were the product of consultants hired by the utilities, did not assure safety and pulled out of the process to take time to develop its own plans.[9]

FEMA's report on the second exercise once more found "significant deficiencies." FEMA concluded: "As of this date, FEMA cannot assure that the public health and safety can be protected in the 10-mile emergency planning zone (EPZ) around Indian Point."[10] Many of the problems identified were the same as those found two years earlier; for example, bus drivers, school officials, and county personnel required to implement the plan had not been trained, and transportation plans were inadequate. Throughout these repeated tests and reviews, Westchester County, Rockland County, the New York State attorney general, the New York City Council, the speaker of the New York Assembly, and numerous others asked the NRC to shut down Indian Point until adequate emergency plans were in place.

In May 1983, after FEMA found Indian Point's emergency plans deficient for the third time, the NRC still did not act to shut the plants down. Instead, it wound the clock back another 30 days, giving Con Ed and PASNY until June 9 either to comply with the emergency planning regulations or to demonstrate that they had "adequate compensating measures."[11]

On the eve of the deadline, FEMA wrote the NRC that "substantial progress" was being made, based on an agreement by New York State to provide "compensating measures" by substituting its personnel for those of Rockland County and a promise by Con Ed and PASNY to train their own personnel to drive buses in place of local drivers. While FEMA made no finding that emergency planning was adequate, it reported that "subject to further evaluation from upcoming tests and exercises, it now appears that continuation of this commitment and momentum should bring about responsive corrections to the deficiencies noted in our earlier report."[12] In other words, for the fourth time Indian Point's emergency plans still failed to meet the regulations, but FEMA hoped that "momentum" would eventually bring compliance.

On June 10, the Commission issued an order (Commissioners Victor Gilinsky and James Asselstine dissenting) not to shut down the plant. The order did not find that the NRC's regulations were met or that emergency planning was adequate at Indian Point. Instead, it found that "adequate compensating actions" had been taken or were "planned to be taken promptly" and that "the level of licensees' involvement in the correction of deficiences . . . has improved significantly."[13] The Commission put forth a novel rationale for having allowed the plant to pass a deadline four times without meeting the emergency planning rules: "Neither the law nor our regulations dictate how many opportunities a licensee has to bring itself into compliance with our emergency planning rules."[14]

The Commission's decision not only left Indian Point with inadequate emergency plans; it also had potentially worse effects by sending the message to utilities all over the country that they need not comply with the NRC's emergency planning regulations either. Commissioner Asselstine charged that the decision "makes a mockery" of the NRC's emergency planning rules.[15] Commissioner Gilinsky summed up the situation this way: "The question before the Commission was a simple one: Was it going to enforce its regulations on emergency preparedness? The answer that emerges is that the NRC will settle for the 'moral equivalent' of compliance."[16]

Fracture Resistance of Reactor Vessels

The possibility of a serious accident resulting from a combination of high pressure, "thermal shock" (a rapid cooling of the reactor vessel), and a vessel embrittled by years of neutron bombardment is a major unresolved safety issue. Having established rules to ameliorate this hazard, the NRC then changed the rules to allow plants not able to meet the standard to continue operating.

In a pressurized water reactor (PWR), the uranium fuel core is held by the reactor pressure vessel, which is 40 to 50 feet tall and fifteen feet in diameter. The eight-inch-thick steel walls of the vessel can lose their ductility, becoming brittle and subject to fracture due to long-term exposure to radiation. While emergency core cooling systems (ECCS) are designed to prevent core melting as a result of a break in the pipes connected to the reactor vessel, the ECCS cannot cope with breaks in the vessel itself.

If an accident occurs, such as a small break in the reactor cooling system or a stuck-open valve, sudden severe overcooling (by cold-water flooding) of the reactor pressure vessel can result, followed by a rapid rise in pressure. The combination of low temperature and high pressure

is referred to as pressurized thermal shock, an unresolved safety issue
that was scheduled for final resolution in December 1985.* The danger
posed by pressurized thermal shock is that small flaws in the vessel can
propagate into a through-the-wall crack and the vessel can rupture. Rupture can occur if the vessel is pressurized while its temperature is below
the "nil ductility transition temperature," a temperature range over
which the characteristics of the vessel change from ductile to brittle. This
transition temperature is very low for a new reactor vessel; for fracture
to be likely, a new vessel must be cooled below about 40° F, far below
the normal operating temperature of approximately 550° F. However,
as a result of long-term exposure to radiation during reactor operation,
fracture resistance decreases and the transition temperature below which
pressurized thermal shock poses a hazard gradually increases.

NRC regulations require that if a vessel's transition temperature is
predicted to exceed 200° F at the end of the plants's lifetime, the reactor
vessel "must be designed to permit a thermal annealing [heat treatment]
process" in order to "recover material toughness properties."[17] Over the
past few years, the NRC has discovered that reactor vessels are degrading
faster than predicted and that if events that have already occurred at
some operating plants were repeated at plants with older vessels, these
vessels might rupture. In July 1981, the NRC ordered 44 operating
plants to perform tests to determine the condition of their reactor vessels.
The results showed that fifteen of these plants had *already* exceeded the
200° F transition temperature that the regulations said should not be
reached before the end of the plant's lifetime without making prior
provisions for thermal annealing.[18]

The NRC did not order these fifteen plants shut down to perform
thermal annealing. Instead, in February 1984 the agency published a
proposed rule that would establish a new transition temperature "screening criterion" of 270° F or 300° F for different parts of the vessel.[19] (The
Commission proposed these higher temperature limits despite an NRC
contractor's report that said it would be "unwise" to do so. Commissioners
Gilinsky and Asselstine noted that this report "tabulates numerous conservative factors, unconservative factors and 'unknown factors' in the
[pressurized thermal shock] analyses."[20]) The NRC staff had previously
informed the Commission that the fifteen plants had transition temperatures that would not exceed the proposed new limits for at least three
more years of operation.[21] The NRC portrayed the proposed rule

---

*The NRC identified it as a generic safety issue as early as 1975.

changes as "intended, if adopted, to produce an improvement in the safety of PWR vessels," not as what they were: a relaxation of existing safety requirements.[22]

The Commission also specifically requested comments on the merits of eliminating the thermal annealing requirements.[23] In 1976, when responding to charges that the potential for reactor vessel rupture had serious safety implications, the agency had stated: "In place annealing of reactor vessels has been demonstrated to be an effective means of restoring the material properties [of the reactor vessel]."[24] In contrast, the Commission's 1984 notice of its proposed rule contained the following factual statement: "Thermal annealing has *never* been attempted on a commercial reactor, let alone shown to be practical."[25]

It is UCS's view that the proposal for higher temperature limitations was made, in large part, because retaining the original limitation would require either permanently shutting down several reactors or undertaking the impractical, if not impossible, thermal annealing process. The Commission's apparent attitude that regulations should be changed to match the condition of the reactors before reactors are changed to comply with the regulations is epitomized by the agency's rationale for the proposed rule: "Nothing undermines respect for the Commission's rules so much as requirements which cannot be met and from which licensees are subsequently exempted."[26]

Operator Unqualification

In the spring of 1984, the Commission discovered that the staff had been subverting an agency regulation that required persons applying for reactor operator licenses at a new plant to have "extensive actual operating experience at a comparable reactor."[27] More important, the Commission probably saw that it would have to disregard the same regulation in order to approve three new plants. To deal with the potential problems raised by this situation, the NRC simply changed the requirement so that actual operating experience was no longer necessary. In fact, it changed the requirement on the very day it was to vote on an operating license for Diablo Canyon, the first of the three new plants whose operator makeup lacked even a remote resemblance to that necessary to meet the regulation.

If the NRC's rule had been enforced, utilities would have had to employ experienced reactor operators to obtain an operating license for a new plant. However, since at least 1976, the NRC staff, without Commission approval to change the rule, had interpreted "extensive actual operating experience" to include "participation in training programs that utilize nuclear power plant simulators."[28] The commissioners were evidently surprised to find out that at the three new plants—Diablo Canyon, Grand Gulf, and Shoreham—none of the operators had "any actual experience operating comparable reactors at full power."[29]

Commissioner Gilinsky asked the NRC's Office of General Counsel for a legal opinion on the proper interpretation of the regulation. Assistant General Counsel James Fitzgerald found that the staff's interpretation "seems to be inconsistent with the plain meaning of the regulation, which calls for *actual*, not simulator, operating experience." Fitzgerald noted: "No provision is made for simulator experience, equivalent experience, or comparable experience."[30]

Chairman Nunzio Palladino then asked the general counsel and the executive director for operations to reconsider the matter in light of the staff's longtime practice. General Counsel Herzel Plaine advised the commissioners that the NRC would be "on dubious legal ground if it asserts that the regulation, as currently written, is consistent with present staff practice." He suggested that the Commission instead change the regulation to match past practice.[31] The staff, while denying that the illegality of their practice was so certain, also favored changing the regulation.[32] Ironically, in another rulemaking proposal dealing with operator training, the staff was taking the position that each shift should have at last one experienced operator.[33] But that was for the future; for the purpose of licensing reactors in the present, the Commission was content to allow a new plant to operate without a single operator with experience at a comparable reactor.

Gilinsky suggested that the Commission at the very least should require that a minimum of one operator on each operating shift have at least one year's experience as a licensed operator at a comparable facility. The Commission majority objected that this requirement would be too difficult; it was by this time known that some new plants could not meet even this relaxed standard. Instead, the Commission decided to have "advisers" on hand who had previously held reactor operator licenses. The question then arose as to who would test the advisers, since an adviser who did not know a particular plant could give the wrong advice and thus be worse than no adviser at all. Gilinsky proposed that they be tested by the NRC. When that met objection, he suggested using the Institute for Nuclear Power Operations, an industry group formed after the TMI accident to foster industry self-regulation. A majority of the commissioners voted to rely instead on the utilities themselves to do the testing.[34]

The staff, the Commission, and its general counsel all argued for eliminating the "actual operating experience" requirement on the premise that this regulation was written before simulators had become so sophisticated. Yet Gilinsky pointed out that reactors had also become more complex. Although simulators are extremely valuable training tools, they cannot possibly provide the equivalent of actual operating experience. Even today's simulators can simulate only a fraction of plant operations. In comparison, simulators used to train airline pilots are far more sophisticated than those for reactors, yet many hours of actual flight time are still required for a pilot's license. As Gilinsky said: "No one would dream of allowing an aircraft to take off with a new crew that had only simulator training."[35]

Furthermore, simulators replicate only the control room; they cannot teach the shift supervisor the job of managing an entire plant. This job includes the responsibility for performing reviews of maintenance and testing procedures critical to the safe operation of the plant and the authority to change accident recovery procedures or to disable safety equipment if necessary.[36] The Kemeny commission found that the Babcock & Wilcox simulator was "not programmed to reproduce the conditions that confronted the operators during the [TMI] accident."[37] The same inadequacy applies to all existing simulators that do not fully duplicate both the physical layout and the operating behavior of the plants for which they are used. The operators at Grand Gulf, Shoreham, and Diablo Canyon were trained on simulators that are not identical to the plants they are licensed to operate.[38]

Gilinsky objected to the assertions by the Commission and the staff that the rule was being changed to be consistent with long-standing agency practice. He pointed out that the sense of the rule—that at least a fair number of operators at a plant would have this experience—had been satisfied until recently, when the Commission had allowed "completely green crews" to start up a plant.[39] Nevertheless, the Commission asked the staff to prepare a proposed rule to allow simulator training to be counted as extensive actual operating experience. It also decided that all previously issued licenses would remain valid and that all new licenses issued before the new rule became effective would automatically be exempted from the old rule. In making these decisions, the Commission ignored Gilinsky's observation that a blanket exemption is essentially a rulemaking and that the Administrative Procedure Act lays out a legitimate way to carry that out, including a requirement that public comment be allowed and considered. After the vote, Chairman Palladino thanked the staff for attending to these issues "in the time frame we had to work in."[40] That same day, Diablo Canyon, without a single reactor operator having "extensive actual operating experience at a comparable reactor," was granted a low-power license.

Ignoring an Old and Inventing a New Regulation at Shoreham

The unreliability of Transamerica Delaval Incorporated (TDI) diesel generators in a number of plants awaiting licenses led to a series of decisions in which the NRC displayed an attitude of disregard for its own regulations. The situation resulted in a policy change loosening the agency's standards for granting exemptions from its rules.

The Shoreham nuclear power plant was unable to obtain a low-power license because its TDI diesel generators, the emergency on-site power source, failed during testing in August 1983 and were found to have dozens of defects, including cracked crankshafts, pistons, cylinder heads, and engine blocks. Attention then was focused on malfunctions that had occurred in TDI engines around the country and on a quality-assurance breakdown at the TDI company.[41] In a January 1984 meeting to discuss resolving these problems, NRC Director of Regulation Harold Denton told industry officials that "we are not prepared to go forth and recommend the issuance of new licenses on any plant that has [TDI] diesels until the issues that are raised here today are adequately addressed."[42] Darrell Eisenhut, the director of licensing, added that this policy would apply "prior to licensing, even a low power license. . . ."[43]

The following month, after the Shoreham owner, Long Island Lighting Company (LILCO), told the licensing board that Shoreham was ready for low-power operation, the NRC staff disagreed with LILCO, maintaining that a variety of questions about the design, manufacture, and quality-assurance process for the diesel generators must be resolved, even for low-power operation. The staff also noted that "General Design Criteria 17 requires an independent, redundant and reliable source of on-site power" and that it took "no position" on whether LILCO could "support an application for an exemption to allow it to go to low-power absent reliable safety-grade diesels."[44]

In March, Chairman Palladino met with the NRC's executive director, executive legal director, general counsel, and other staff personnel to discuss ways to expedite the Shoreham low-power license. Immediately after this meeting, the staff changed its position and supported low-power operation before the TDI issues were resolved and without an exemption from GDC 17. Instead, in what Commissioner Gilinsky described as "concocting arguments on how all this can be rationalized,"[45] the staff proposed that LILCO be allowed to run at low power without the diesel generators if public protection was equivalent to or greater than it would be at full power with the approved diesel generators.[46] Gilinsky later characterized the staff's actions as "trying to run legal interference for the Company."[47]

The licensing board adopted the NRC staff's new position. Three commissioners—Palladino, Thomas Roberts, and Frederick Bernthal—voted to let the board's decision stand (Gilinsky and Asselstine dissented),[48] in spite of the finding of the general counsel that "the Commission's regulations do not generally distinguish between low and full power. As a result, with very limited exceptions, all of the Commission's regulations . . . applicable to onsite power supply (General Design Criteria 2, 4, 5, 17, 18, and 50) are applicable to low power to the extent they are technically relevant."[49]

The counsel for Suffolk County (New York), an intervenor in the Shoreham proceeding, argued to the Commission that LILCO should be required to meet the standards for an exemption from GDC 17:

> The reason for this sensible approach, and it's common in the regulations of virtually every regulatory agency in this city, perhaps in this country, is that regulations must have integrity. They must be understood, they must be certain. And the only way in which there can be a deviation from those has to be through a procedure which puts parties on notice and gives them a right to litigate the implications of changes.[50]

Suffolk County further argued that the staff's standard amounted to the creation of a new regulation "custom tailored" to LILCO's inability to meet GDC 17.[51] Eventually, after a public outcry and a congressional hearing on the NRC's actions, the Commission reversed itself and said LILCO should request an exemption.[52]

Double Standards at Grand Gulf

Grand Gulf also had TDI diesel generators that were implicitly subject to Denton's January 1984 statement that low-power operation would not be permitted until the TDI problems were resolved. In late 1983, Grand Gulf was shut down while the utility was performing required retraining and recertification of its operators. Yet in early May, Grand Gulf resumed low-power operation with no demonstrably reliable on-site emergency power source.

Later that month, in response to an inquiry from Commissioner Gilinsky, the NRC staff reported that Grand Gulf did not meet GDC 17 requirements but was being permitted to operate anyway because in the staff's view it posed no undue health and safety risk. Gilinsky said that was "precisely the argument the Commission rejected in the more heavily publicized Shoreham case."[53]

To make matters worse, less than one week after the Commission made clear that GDC 17 applied to low-power operation at Shoreham, the NRC staff issued an order allowing Grand Gulf to operate at low power for weeks while one TDI diesel generator was taken out of service for inspection,[54] diminishing the plant's on-site emergency power supply. The staff's order changed the plant's licensing conditions, which would otherwise have required plant shutdown. The owner, Mississippi Power and Light (MP&L), was to file a request for an exemption from GDC 17, but the staff allowed operation to continue while the request was processed.[55]

The NRC staff and its lawyers came up with a number of explanations for their order. For example, they argued that they did it to *improve safety*; ordering the diesel to be torn down would result in "increased assurance as to reliable onsite power" during low-power operation.[56] That might imply that the staff believed there was a reason to have reliable on-site power at low power. Yet they had been allowing the plant to operate with unreliable diesel generators for almost a month, and the new order had relaxed the licensing conditions even further by allowing one diesel generator to be completely disabled. In an internal memo, the NRC's general counsel advised the Commission that the legal basis for the staff's order was "questionable," noting that "[i]t follows that there was little or no low power operation safety basis" for ordering the inspection itself, much less with the simultaneous relaxation of the requirements.[57]

At a hearing in June 1984, Representative Richard Ottinger, chairman of a House Energy and Commerce subcommittee, charged the Commission majority with "contempt for law and its own regulations," as illustrated by Commissioner Bernthal's assertion that the Commission had judged the plant safe and then went "parsing and picking at regulations."[58] Commissioner Asselstine also took issue with Bernthal's casual attitude toward NRC regulations:

It seems to me that if you have a body of regulations and . . .you have requirements in a license and a plant for some reason does not meet those, then our regulations and our whole regulatory process lays out an orderly way to examine the health and safety questions of whether in that situation a plant ought to be permitted to continue to operate.

I think the key question here is the approach that the staff has used. In essence, I think, what the staff has done is said, "We can make a back-of-the-envelope estimate of what the health and safety risk is here, and because we think it's low enough, we can use a very different approach," an approach that is not—I don't believe—envisioned either under the Atomic Energy Act or under our regulations, to reach the conclusion that we want to reach, which is to keep this plant running.[59]

Gilinsky also defended the agency's general design criteria as a keystone of nuclear safety:

Let me say, the general design criteria reflect a great deal of thinking in the Commission years back, a great deal of work. It has gone through a process of review. Those are the basic requirements of the Commission.

If you think they are wrong, perhaps we ought to change them. But let's propose a change. But to just casually set them aside and say they can otherwise be casually set aside, I think is to set a terrible precedent.[60]

Most conspicuous was the fact that the agonizing process the Commission had just gone through in its attempt to sweep aside GDC 17 at Shoreham had led it to take a position directly contradictory to what it had done with Grand Gulf. To that the NRC legal staff responded that the two situations were different—Grand Gulf already had a low-power license and the question was whether to take it away, whereas Shoreham did not have a license and the question was whether to grant it one.[61] The significance of this distinction is not clear; in both cases, the issue was whether a plant can be operated safely despite noncompliance with well-established safety requirements. The NRC has always asserted the authority to take immediate action when there is a question about the safety of licensed plants. Moreover, the staff's adoption of this argument is ironic in light of its specific request during the Shoreham deliberations that the Commission determine the applicability of GDC 17 to low-power operation, precisely because the issue would apply to other plants, including Grand Gulf.[62]

In fact, the most important difference between the two cases was that with Shoreham there was forceful opposition from intervenors, whereas with Grand Gulf there were no intervenors to challenge the agency's action.

Standards for Exemptions from Safety Regulations

In its May 16, 1984, Shoreham decision, the NRC ruled that LILCO must request an exemption from GDC 17. To obtain an exemption, the Commission said, LILCO must show that "exigent circumstances" favored granting one and that operation of the plant without a fully qualified on-site emergency power source would be as safe as operation with it.[63] In other words, operation in violation of GDC 17 must be "as safe as" operation in compliance. Once the controversy surrounding the NRC's actions in the Shoreham case died down, however, the agency quickly retreated from this standard.

Two months after the Shoreham decision, the NRC staff told the Commission that applying the Shoreham standard would interfere with the licensing of other new plants and with the operation of others. According to the staff, the Shoreham standard differed significantly from its past view of the regulations as being "reasonably flexible," when the staff often allowed rules to be met after various periods of time rather than requiring licensees to file exemptions. Even when exemptions were required, the staff, in deciding to grant them, would judge that there was "no undue risk" rather than that the exemption left the plant "as safe as" it was without it.[64] Saying that it was only asking for guidance and was "not going to try to sell the Commission on one point of view or the other,"[65] the staff subsequently prodded the commissioners for permission to return to business as usual.

At a July 1984 meeting, the staff again raised the specter of licensing delays and plant shutdowns. The NRC's deputy director for regulation, Edson Case, said the agency had granted 80 exemptions from its regulations over the previous six months and even if the Shoreham standard were relaxed so that a plant need only be "*substantially* as safe as" it would have been if it complied with NRC rules, one plant shutdown per month would result.[66] Case also warned that four plants soon to be ready for operating licenses would need between five and ten exemptions in order to run at low power. He conceded that relaxing the standard to "substantially as safe as" would take care of 90 percent of the problem, but

> If there is one exemption that can't be granted under the as-safe-as standard, then the plant startup, using our previous way of doing business, will be delayed.
>
> I would estimate that it might be two to three months' delay for [new licenses]. And in particular, for one of the two that you're going to discuss next week.[67]

The Commission and the staff debated various ways of avoiding this stricter standard. Chairman Palladino offered the most novel interpretation of the Shoreham standard: apply it only to the general design criteria and no other regulations.* Yet Palladino could offer no logic for this interpretation:

> Commissioner Asselstine: My difficulty . . . is if you begin to try to parse things out, where do you draw the line? Why is compliance with the general design criteria so important that you're going to apply the Shoreham standard to those but not Appendix A to Part 100, for example?
>
> Chairman Palladino: Because that's what the Commission did.
>
> Commissioner Asselstine: . . . I think there has to be a well thought out, logical, rational basis for drawing that distinction. It has to be more than, well, "The commission was focusing on the [GDC], and that's all we were focusing on."
>
> Chairman Palladino: And that is true; that is all we were focusing on.[68]

Taking Palladino's interpretation one step further, Commissioner Lando Zech proposed that the Shoreham standard "apply to that case only. . . ."[69] and advocated going back to the staff's old way of doing things. William Dircks, the executive director for operations, rallied in support of this idea.

In response to Asselstine's suggestion that the Commission take just a few days to put more thought into their guidance, Dircks objected that "a couple of days is even costly."[70] Case later revealed that Grand Gulf would need exemptions from several general design criteria to obtain a full-power license the following week, and the staff was worried that it could not even establish that the reactor's operation would be "substantially as safe as" if it were in compliance. Asselstine and Bernthal appeared surprised to learn of this state of affairs:

> Commissioner Asselstine: But even with substantially as-safe-as, you can't make that judgment?
>
> Mr. Case: I just don't know, Jim. I would be worried about it.
>
> Commissioner Bernthal: The staff has to—I'm—we don't want to get into the Grand Gulf case here today, but don't you normally have to make the judgments, substantially as-safe-as?
>
> Mr. Case: No.

---

*The general design criteria in Appendix A to 10 CFR Part 50 establish the "minimum requirements" for the design of those structures, systems, and components that are relied upon to provide "reasonable assurance" that the plant will not pose "undue risk to the health and safety of the public." Other regulations have no less importance in terms of public safety. For example, 10 CFR Part 100 establishes the maximum allowable radiation exposure to plant workers and the public following a postulated accident.

> Commissioner Bernthal: We're not even saying as-safe-as; we're say-
> ing substantially as-safe-as.
> Commissioner Roberts: No.[71]

Thus, not only would the NRC allow reactors to operate that did
not comply with its regulation defining the minimum level of safety, but
also these reactors would not even have to be substantially as safe as if
they met the regulation.

The staff also revealed its concern that exemptions granted for Diablo Canyon's low-power license, which were soon to be carried over to
the full-power license, would run into trouble. In response, Bernthal
suggested that they apply the "substantially as safe as" standard only to
operating plants and "simply make a judgment on" those plants up for
operating licenses soon.[72] Finally, Chairman Palladino and Commissioners Bernthal, Roberts, and Zech opted for Zech's proposal. The Shoreham standard would be ignored altogether and the staff could continue
to grant exemptions that could leave reactors not even substantially as
safe as they would be without these exemptions. The Commission would
give additional thought to a long-term policy. (In September 1984, the
staff presented its proposed long-term policy, which would essentially
codify its past practice.[73])

Deputy General Counsel Martin Malsch advised the Commission
that "it would be helpful" if the staff would make the "substantially as
safe as" finding whenever it could. He said that while he thought what
the Commission was doing was lawful, "there may be some difficulties
in explaining that kind of a thing" if it were to be challenged:

> just speaking in common sense, if someone should look at a plant and say,
> "Well, this plant [sic] has been granted, it's not in compliance with Com-
> mission regulations." Okay. That happens. . . . The Commission can't an-
> ticipate every conceivable circumstance. But to say not only is it not in
> compliance but that it's not even substantially as safe as it would be if it
> were, that's something entirely different.[74]

Palladino told the staff to regard that as an "admonition" from the general counsel, not a "Commission-endorsed matter."[75]

Ignoring Earthquakes and Regulations at Diablo Canyon

The Atomic Energy Act and NRC regulations require that if an
issue is material to a licensing decision, then an opportunity for a hearing
on the issue must be granted. This requirement was reinforced when

UCS sued the NRC for adopting a rule that would have prohibited licensing boards reviewing the adequacy of emergency planning in operating license hearings from considering the results of emergency planning exercises. In rejecting the rule as illegal, a U.S. Appeals Court ruled in May 1984 that "because the rule denies a right to a hearing on a material factor relied upon by the Commission in making its licensing decisions, the rule was issued in excess of the Commission's authority under section 189(a). . . ."[76] The Supreme Court refused to accept an appeal of this decision. Nevertheless, at both San Onofre and Diablo Canyon, the NRC refused to allow licensing boards to review the potential effect of an earthquake on emergency response during an accident. In the San Onofre case in 1981, the Commission put off resolving this important issue by labeling it generic and promising a generic rulemaking. By 1984, this rulemaking had not taken place, and the issue again came before the Commission at Diablo Canyon. In denying the intervenors a hearing on the issue, the Commission argued that there was a "low probability" of an earthquake disrupting emergency planning.

In a lengthy dissent, Commissioner Asselstine detailed the flaws in the Commission's reasoning, concluding: "The Commission's decision ignores fundamental principles of emergency planning, offends common sense, and abuses the legal process."[77] Asselstine reminded the other commissioners that the purpose of the NRC's emergency planning regulations was to be prepared for low-probability events; no matter how safe the agency tried to make nuclear plants, there was always the possibility that an unforeseen event would occur—an outlook forced on the Commission by Three Mile Island. In fact, said Asselstine, the "probability arguments used by the Commission are really arguments that we do not need any emergency planning."[78] Yet all major studies of the TMI-2 accident, congressional actions, and the Commission's regulations maintained that emergency planning was needed.

Asselstine stated that 90 percent of the seismic activity in the United States occurs in California and that the NRC had already recognized the uniquely high seismicity of this area by requiring Diablo Canyon to be designed to withstand ground motion almost twice that demanded of plants in other states. He also pointed out that the NRC had taken into account the effect on emergency planning of such other natural phenomena as tornados in the Midwest and hurricanes in certain coastal areas, including the Diablo Canyon site. Yet these phenomena were not shown to be any more likely than earthquakes in California.[79]

Moreover, Asselstine continued, the Commission had denied the intervenors a hearing on the basis that the issue was not material to a licensing decision, although its own technical experts had repeatedly indicated that it was. In June 1982, the NRC staff had stated that it did not believe a generic rulemaking was necessary because earthquakes were very unlikely in most parts of the country. However, the staff saw California as a special case:

> Planning for earthquakes which might have implications for response actions or initiate occurrences of the [more serious] classes in areas where the seismic risk of earthquakes to offsite structures is relatively high may be appropriate (e.g. for California sites and other areas of relatively high seismic hazard in the Western U.S.).[80]

Asselstine pointed out that the staff, asked later by the Commission to clarify its position, had repeated this assertion.[81] The staff had also told the Commission in June 1982 that it requests California utilities and the Federal Emergency Management Agency to "consider" the effects of earthquakes in their emergency planning and review.[82] Asselstine also noted that in the spring of 1984, as the Commission was deliberating on Diablo Canyon's full-power license, it became apparent that the staff's position might require allowing a public hearing on the adequacy of this aspect of the emergency plans before issuing a license. Asselstine said that the staff then "attempted to reverse course," explaining that its reviews of this issue were "informal."[83] Yet, Asselstine pointed out, if an issue is recognized to be material to a licensing decision, which the staff conceded by requiring applicants to consider it, an "informal" review cannot legally substitute for a more rigorous, formal one.[84]

According to a letter to Palladino from Representative Ottinger, Commission meetings that were closed to the public and confidential documents both showed that "the general counsel advised the Commission that the issue of the complicating effects of an earthquake upon the emergency response plan is most probably a material issue, entitled to an adjudicatory hearing prior to the issuance of the license, under Section 189(a) of the Atomic Energy Act."[85]

Ottinger's letter also revealed that in a July 1984 confidential memo, General Counsel Herzel Plaine had told the Commission that "there is 'no convincing, rational basis for the Commission's view that the complicating effects of earthquakes on emergency response deserves [sic] no consideration.' "[86] Similarly, in a closed meeting, another of the Commission's counsel had warned of "the absence of any material on the record supporting the Commission's view."[87]

Despite these cautions, the NRC granted Diablo Canyon a license with no hearing held on this outstanding issue. Moreover, the agency presented its case to a U.S. Appeals Court in a brief that Ottinger concluded "seriously misstates the Commission's actions on the case." The court had issued a stay on Diablo Canyon's ascent to full power on August 17, 1984, partly as a result of Asselstine's objections.[88] Ottinger came to this conclusion after reviewing the Commission's secret deliberations on the court case—material the court had not seen when it lifted the stay in late October. Ottinger subsequently asked the Department of Justice to investigate the NRC's statements to the court.[89]

Ottinger determined from his review that the Commission's decision to license the plant relied on off-the-record material and that its decision not to subject the issue to hearings was made primarily to avoid delay in full-power operation of the plant.[90] The Commission majority staunchly opposed making these deliberations public. Ottinger wrote to Palladino urging that the Commission end this resistance:

> I can only infer . . . that the Commission's interest in withholding these documents and transcripts from public disclosure is motivated by a desire to conceal the Commission's irregular and improper consideration of these matters, and not by an interest in preserving the integrity of the Commission's decision-making process.[91]

The Commission again refused to make the secret deliberations public. It argued that the public interest would be better served by keeping them private, because in discussing legal proceedings in the future the knowledge that discussions would remain private would allow the Commission and staff to be candid.[92] Commissioner Asselstine disagreed:

> I believe that there is a broader, and overriding, public interest in this case which calls for the public release of the transcripts of the Commission's deliberations on the question of the complicating effects of earthquakes on emergency planning. . . . This is the public interest in identifying and correcting serious abuses by the Commission in the conduct of its adjudicatory proceeding, and taking steps to assure that similar abuses do not recur in the future.[93]

The transcripts of the Commission's closed meetings were subsequently leaked. In addition to fully confirming both Asselstine's and Ottinger's statements, they provide a rare glimpse of the commissioners when they are not speaking for the public record.

The context was provided at the outset by the NRC's general counsel:

> Mr. Plaine: Thank you, Mr. Chairman. Our paper suggests that as a result of having waited so long to deal with this issue, we are now down to the time when the full power licensing is about to be considered without any

clear resolution of this issue of the effects of earthquakes on emergency planning in connection with full power operation.[94]

Plaine's assistant then informed the Commission that "[e]verywhere else, site-specific natural phenomena" have been considered in the review of emergency planning, and stated the Commission's dilemma more starkly:

> So, this becomes the unique case. It's these considerations which I think have led us primarily to find some way to proceed with the licensing of this plant but not to have the Commission look as if it is ignoring this problem.
> Chairman Palladino: Okay.[95]

The Commission proceeded, at great length, to explore every possible stratagem to issue the Diablo Canyon license without being reversed in court. Each method discussed was conceded to be of dubious legality, primarily because of the lack of facts in the record to support a safety finding.

The general counsel suggested an approach both sensible and arguably legal:

> Mr. Plaine: . . . if we are trying to give a hearing, don't make it look so bad that a court will say, "My God, what kind of a hearing are you inventing now." We have rules for hearings under the statute itself and we are trying to move into the area of giving a hearing without necessarily making it pre-licensing if we can avoid part of it from being—making it whole [sic] between licensing. There is a good chance that if you actually signed an order now that called for a hearing and a Hearing Board was assembled, it might very well get on with this problem while you are still fiddling with some other questions that will arise with regard to Diablo Canyon.
> Commissioner Asselstine: Which is what the commission should do.[96]

This approach, which would have resolved the safety issue but might have held up the license, was simply unacceptable to the Commission majority and received no serious consideration. Commissioner Roberts's earlier question is revealing:

> May I ask Sheldon a question? If you were of the view that it would be unfair to the licensee at this point to deny a licensing of the plant, assuming all the other requirements are necessary, and you are only concerned with winning the case, which alternative would you choose?
> (Laughter)[97]

In short, the transcripts provided rare evidence that the Commission's preoccupation with getting plants into operation had become the overriding driving force at the agency, sweeping questions of legal rights and even of safety into the background.

Ignoring Congress: The Sholly Amendment and the Sholly Rule

The Atomic Energy Act requires the NRC, when amending a construction permit or operating license, to "grant a hearing upon the request of any person whose interest may be affected by the proceeding. . . ."[98] Nevertheless, in 1980 the NRC allowed the operator of the Three Mile Island plant, General Public Utilities, to release radioactive krypton gas trapped inside the reactor building of the damaged plant but did not allow an opportunity for a prior hearing.[99] Two Pennsylvania residents, Steven Sholly and Don Hossler, sued the NRC, and in November 1980 a U.S. Court of Appeals ruled that the agency's action was illegal.[100] The NRC went to Congress, objecting that the court's ruling would require a hearing to be held before even the most routine license amendments that were "insignificant from a public health and safety standpoint" and where there were "no new safety issues raised, no unreviewed hazards" or "safety-connected issues."[101] In 1982, in response to the NRC's concerns, Congress enacted an amendment to the Atomic Energy Act that became known as the Sholly amendment. It allowed the NRC to amend operating licenses before holding a requested public hearing only if the amendment involved "no significant hazards consideration."[102]

In granting the Commission's request that it be permitted to make minor license amendments effective before holding a requested hearing, Congress was sensitive to the potential for abuse of the "no significant hazards consideration" criterion. It therefore required the NRC to develop guidelines that "draw a clear distinction" between license amendments that do and do not pose significant hazards considerations. Congress further required that the standards be "capable of being applied with ease and certainty."[103] Congress also directed that the NRC's standards "ensure that the NRC staff does not resolve doubtful or borderline cases with a finding of no significant hazards consideration."[104]

Furthermore, Congress made it clear that the NRC's standards must only require a decision on whether a license amendment posed significant hazards *considerations*, not whether it posed significant hazards. Thus the decision on whether to permit a hearing should only concern whether the amendment involves significant safety-related questions. As the House-Senate Conference Committee stated: "These standards should not require the NRC staff to prejudge the merits of the issues raised by a proposed license amendment. Rather, they should only require the staff to identify those issues and determine whether they involve significant health, safety or environmental *considerations*."[105]

In May 1983, the NRC passed a rule implementing the Sholly amendment that strayed far beyond the boundaries set by Congress. This "Sholly rule" did not draw a clear distinction between license amendments that pose significant hazards considerations and those that do not, nor was it capable of being applied with ease and certainty. Rather, the NRC's rule consisted of unlimited and undefined standards that delegated virtually complete discretion to the NRC staff: The rule specified that the staff could make a "no significant hazards consideration" determination if it found that the amendment would not involve a "significant increase" in the probability and consequences of accidents previously evaluated, would not create the possibility of a "new or different kind of accident," and would not involve a "significant reduction" in a margin of safety.[106] By the nature and complexity of the questions they pose, the rule's standards force the staff to undertake detailed analysis that, while necessary for a decision on whether the amendment should be granted, is neither necessary nor lawful for a determination of whether there should be an opportunity for a hearing before the amendment is effective.

The answers to these complex questions are generally not apparent on the face of a license amendment; rather, these determinations require the use of methods such as probabilistic risk assessment (PRA), which cannot be applied with "ease" and are clearly incapable of being applied with "certainty."* As a result, the NRC's determinations under the Sholly amendment are extremely prone to misuse. Even if PRA were capable of yielding a reasonably accurate and objective answer, its use goes far beyond the identification of issues that Congress envisioned in enacting the Sholly amendment. If PRA is used in decisions made before a hearing, its inherently unreliable assumptions and conclusions are not subject to any independent scrutiny (which the hearing would have otherwise provided). Under the Commission's rule, the staff decides during the preliminary finding the same issues that determine whether the amendment should be approved *at all*. That is precisely what Congress stated the NRC was *not* to do.

The NRC also declined to include in the rule two proposed lists of types of license amendments likely and not likely to involve significant hazards considerations. Such lists would have given some substance to the rule's standards and helped assure that the rule could be applied with "ease and certainty," as Congress intended. Instead, these examples

---

*See chapter 2, section on "Cost-Benefit Analysis."

were listed in the preamble to the rule, where, as Commissioner Gilinsky observed, "they will be of little or no legal consequence."[107] Important examples were left out, and many of those included were vague or repeatedly used the word "significant," leaving critical decisions to the judgment of the staff.

Before the passage of the Sholly amendment, the NRC's practice was to approve all but the most exceptional license amendments before offering an opportunity for a hearing. Congress was aware of this practice and, in directing that "the Commission will use this authority carefully,"[108] told the Commission to change its ways. The standards the Commission put in its Sholly rule and the staff's handling of license amendments since this rule went into effect indicate that agency practices have not changed. Out of 1,415 license amendments proposed under the Sholly rule between its adoption on May 6, 1983, and August 22, 1984, *only 25* were determined to involve significant hazards considerations.[109]

The NRC's claim that over 98 percent of the requested license amendments during that period involved no significant safety issues strains credulity. License amendments involving significant safety issues but which the staff found to involve "no significant hazards considerations" included the following: skipping required tests of instrumentation needed to determine plant status after an accident, changing the list of safety equipment required to be available during plant operation, changing the setpoint of pressure relief valves necessary to prevent rupture of the reactor containment building, changing the means of controlling hydrogen buildup in the reactor building, using different methodology to reanalyze the setpoint for automatic reactor shutdown to prevent fuel damage, and setting new limits on the degraded plant conditions under which operation can continue.[110]

Several other examples of the NRC's misuse of the Sholly amendment are described below.

Steam Generator Repairs at Three Mile Island Unit 1

The misapplication of the Sholly amendment is perhaps best illustrated by the staff's handling of the unprecedented repair techniques used on the steam generators at Three Mile Island Unit 1 (the plant unaffected by the 1979 accident). In 1981, a chemical used in the containment spray system was unintentionally mixed into the reactor coolant system and caused extensive corrosion of the tubes in both steam generators. General Public Utilities (GPU) proposed to repair the steam generators using explosives to expand the tubes, thereby reducing the leakage of reactor cooling water into the steam generators to an acceptable value.

In August 1982, the staff informed GPU that the proposed "kinetic [explosive] expansion repair technique" appeared to involve some "unreviewed safety questions":

> 1. The corrosion mechanism and extent of corrosion in the steam generators are unique. The staff has not reviewed the potential consequences of additional plant operation subsequent to the repair of the defects. In particular, the potential for this type of corrosion to reinitiate during operation and to rapidly progress, adversely affecting the steam generator primary pressure boundary, will need to be reviewed by the staff.
> 2. The potential exists for this type of corrosion to attack other primary system pressure boundary materials.
> 3. The proposed tube repair technique has not previously been approved by the staff as an acceptable method for repairing defective steam generator tubes.[111]

The staff informed GPU that a license amendment therefore would be required to perform the repair and to resume operation.[112] Two months later, the staff reversed itself by finding "that the repair did not involve an unreviewed safety question or a modification to the technical specifications, and hence could be conducted without prior NRC approval." The staff did, however, retain its position that NRC review and approval were required prior to any subsequent power operation of the plant.[113] Thus the repairs were allowed to be performed without a license amendment or a prior public hearing.

GPU completed the repairs and requested a license amendment to allow operation. On May 31, 1983, less than one month after the Sholly rule went into effect, the staff made a proposed finding that operation of the plant posed "no significant hazards consideration." The Federal Register notice stated that none of the examples of amendments either likely to or not likely to involve significant hazards considerations applied to the amendment.[114] But in fact, the amendment fell within example VI of those likely to involve significant hazards considerations: "A change to technical specifications or other NRC approval involving a significant unreviewed safety question."[115] The proposed finding was also justified by the assertion that compensatory measures would be employed to provide a level of safety commensurate with that anticipated if the repairs had not been needed.[116] Three Mile Island Alert, a local citizen group, promptly requested that a hearing on the safety of operation be held before TMI-1 was restarted.

Representative Ottinger called the staff's proposed finding "an outrageous determination and a complete disregard of the provisions that we made last year and of the report language that indicated our intent" in framing the Sholly amendment. He went on:

> it was my clear understanding when we did this last year that this was to take care of routine matters so you wouldn't have to have a hearing on a whole bunch of very inconsequential amendments, it was not to deal with any matter which was of real substance.[117]

Commissioner John Ahearne claimed at Ottinger's authorization hearing that this determination "is going to be a close call . . . initially this sounds like something that would fall outside the boundary of what we would have thought of for Sholly."[118] Even if one accepts the tenuous proposition that this was a "close call," Congress had unambiguously rejected the use of the Sholly authority for borderline cases.

In spite of congressional criticism, the staff made its final determination in November 1983 that the license amendment involved "no significant hazards consideration." Commissioner Asselstine observed:

> [the staff] made its preliminary and final determinations of no significant hazards consideration based on the *merits* of the amendment itself—i.e., on whether the amendment posed significant additional risk to operation of the plant.
>
> Unfortunately, that determination is not the determination called for by the Sholly amendment. Rather, as its legislative history makes abundantly clear, the Sholly provision requires the Commission to determine whether the amendment presents any significant safety questions, i.e., whether the amendment poses any significant new or unreviewed safety issues for consideration.[119]

The NRC's general counsel also cautioned the Commission not to concur with the staff's determination. According to a trade press article, a confidential memo from the general counsel noted: "It appears to us that under the staff approach nearly all operating license amendments issued by the commission would qualify as involving no significant hazards consideration and therefore could be issued without a prior hearing." The general counsel expressed additional concern that the staff was making numerous "no significant hazards consideration" determinations "premised on an interpretation that presents serious legal problems."[120]

In January 1984, Chairman Palladino and Commissioner Roberts nonetheless voted to concur with the staff's "no significant hazards consideration" finding. Commissioners Gilinsky and Asselstine opposed it. Commissioner Bernthal refused to vote, stating that the plant was not ready to operate anyway and that he did not want to make any "unnecessary decisions."[121] Bernthal drew strong criticism from members of the public and Representative Edward Markey. Markey told Bernthal that his decision not to vote had "far-reaching policy ramifications. . . . The significance of the deadlock imposed by your abstention is that the Commission silence offers an implicit endorsement of the staff views. . . ."[122]

Expansion of Spent Fuel Pools

One category of amendments was explicitly singled out by Congress as clearly involving significant hazards considerations: the expansion of storage capacity for spent fuel at each reactor site. This expansion is most often accomplished by "reracking"—replacing existing spent fuel storage racks to permit spent fuel assemblies to be spaced more closely. Before the Sholly amendment was enacted, the NRC routinely offered an opportunity for a hearing before granting a reracking license amendment.[123] Congress made it clear that this practice was to continue.

During the House debate on the bill, Representative Ottinger replied to a question on this subject:[124]

> The expansion of spent fuel pools and the reracking of the spent fuel pools are clearly matters which raise significant hazards considerations, and thus amendments for such purposes could not, under section 11(a), be issued prior to the conduct or completion of any requested hearing or without advance notice.

Similarly, the Senate committee report declared that "the Committee anticipates, for example, that, consistent with prior practice, the Commission's standards would not permit a 'no significant hazards consideration' determination for license amendments to permit reracking of spent fuel pools."[125] The lawyers for both the Commission and the staff flagged this obvious congressional mandate when they wrote in a memo to the Commission that "every reference" from both houses of Congress "reflects an understanding that expansion and reracking of spent fuel pools are matters which involve significant hazards considerations."[126]

Yet in promulgating the Sholly rule, the Commission not only omitted the reracking example from the rule but even removed it from the preamble's list of examples of license amendments likely to involve significant hazards considerations. Instead, the Commission concluded that because some "reracking technology has been well developed and demonstrated," the matter needed further study.[127] Both UCS and the State of Maine (which then had a reracking hearing in progress) objected to the Commission's disregard of clear congressional intent.[128] Utilities and industry trade associations praised the Commission's elimination of any mention of reracking in the proposed rule, and this position prevailed in the final rule. In August 1983, the staff sent its study on reracking under the Sholly rule to the Commission. The study concluded that two rarely used methods of spend fuel pool expansion (requested for only three reactors as of August 1983) would involve significant hazards considerations but that a "no significant hazards consideration" finding could be made for reracking.[129] On the same day, the staff announced that it was making a no significant hazards finding for a spent fuel pool expansion at Oconee, in South Carolina.[130]

License amendments involving reracking may, under many circumstances, pose no undue risk to the public. But such amendments clearly involve significant hazards considerations. To exempt reracking amendments from the law would require action by Congress.

Random Application of the Sholly Rule at Grand Gulf Unit 1

The Atomic Energy Act and the Sholly rule require that, in addition to allowing an opportunity for a hearing on a proposed license amendment, the NRC must grant a 30-day period for public comment on the NRC staff's proposed finding that an amendment involves "no significant hazards consideration." This period may be waived only in "emergency or exigent circumstances."[131] At Grand Gulf 1, the staff followed what appears to be a random course in applying these provisions to license amendments. Between July 1983 and February 1984, the NRC issued six license amendments that made more than 100 changes to the license conditions and technical specifications—plant-specific requirements that reflect the NRC's assessment of which aspects of the plant must be closely controlled in order for the plant's safety evaluation to be valid. Of these 100 changes, the NRC chose to give notice of only ten according to the Sholly procedure, and in only six cases was 30-day advance notification given. Notice of the other four was given on an emergency basis in newspapers.[132]

Even though for 90 changes the NRC gave no advance notice of any kind, the staff, in reporting after the fact that the amendments had been issued, claimed that "prior public notice" of the amendments had been given implicitly on July 28, 1978—the day notice was given of an opportunity for a hearing on the operating license.[133] This notice was published at least five years before the changes were proposed.

The NRC also evaded the law and its own regulations in May 1984, when the NRC staff allowed Grand Gulf 1 to operate with part of its emergency power source inoperable but did not classify this relaxation of the reactor's license conditions as a license amendment. The NRC's general counsel advised the Commission that this action was a license amendment, subject to the public hearing provisions of the Atomic Energy Act and the Sholly amendment. The general counsel recommended requiring the utility to apply for a license amendment, which he believed could be done with no disruption of plant operation.[134] Commissioner Asselstine, the only lawyer on the Commission, also pointed out that "[t]his is precisely the type of situation that the Sholly Amendment was intended to deal with" and later at a congressional hearing characterized the agency's action as "illegal."[135] The Commission majority, however, decided to uphold the staff's decision not to classify the action as a license amendment and thus circumvented the law.

Word Engineering

At Millstone 2, in Connecticut, the operator, Northeast Nuclear Energy Company, requested an amendment to allow, among other changes, reduced coolant flow through the reactor, higher temperatures in the fuel, and continued reactor operation with safety equipment inoperable for a longer period. The utility's safety analysis of these changes predicted an increase in accident consequences—one of the Sholly rule criteria indicating that a license amendment involves significant hazards considerations. However, the staff proposed a "no significant hazards consideration" finding, claiming that the more severe consequences "are within the acceptable limits" incorporated in the operating license. In the case of the Millstone amendment, the proposed finding acknowledged that the increase in consequences resulting from the proposed license amendment resulted in the need for *"detailed Staff review."*[136] That would indicate a far more complex issue than the routine amendments envisioned by the framers of the Sholly amendment.

Moreover, a draft of the NRC staff's proposed "no significant hazards consideration" finding stated that "two unreviewed safety questions are involved."[137] After an NRC staff lawyer pointed out that this statement would place the amendment under one of the examples of amendments that *are* likely to involve significant hazards considerations,[138] mention of unreviewed safety questions was dropped from the staff's final proposed determination.

Frustration of Congressional Oversight Efforts

The Atomic Energy Act, as amended, provides that "the Nuclear Regulatory Commission shall keep the committees of the Senate and the House of Representatives which, under the rules of the Senate and the House, have jurisdiction over the functions of the . . . Commission, fully and currently informed with respect to the activities of the . . . Commission."[139] This responsiveness to Congress is crucial, because congressional oversight is the primary mechanism to keep the Commission, as a presidentially appointed body, accountable to the public. The NRC, however, has on a number of occasions dragged its feet in responding to inquiries and document requests from the congressional subcommittees charged with oversight of the agency, and in some cases it has been plainly uncooperative.

In an April 1982 letter to the Commission, Representative Ottinger wrote:

> in the past several months, there have been a number of instances wherein the Commission not only failed to keep the relevant committees informed, but has withheld, for an unnecessary period of time and without authority, information which should have been submitted to Congress and which had been specifically requested by such committee[s]. The number and nature of these instances form a pattern of behavior on the part of the Commission which is inconsistent with its statutory responsibility and which constitutes a totally unacceptable level of performance on the part of the Commission, its Chairman and members. While the Commission's efforts to withhold or delay the release of information is not unique within the context of this Administration, the number of instances and the length of the delays set the Commission apart.[140]

One instance of delay involved a request for information regarding the NRC's proposed changes in its licensing process. By the time the request was answered—seven months later—the changes had already been made into a final NRC rule. Thus the Commission frustrated the efforts of the subcommittee to express its concern and have an impact on the operations of the Commission.[141]

Ottinger also found that "the delays are then frequently compounded by the fact that the information submitted by the Commission is nonresponsive."[142] For example, on December 18, 1981, the subcommittee informed the Commission of its concern over possible material false statements made to the NRC by Pacific Gas & Electric Company (PG&E). On February 4, 1982, the subcommittee wrote the Commission that the agency's investigation of PG&E was inadequate, citing significant omissions and defects in the report.* According to Ottinger, the Commission responded to the subcommittee's December 1981 letter simply by saying that the subcommittee's concerns "had been fully resolved" by the investigation.[143]

In another case, Commissioner Bernthal wrote to Representative Markey's subcommittee that he was "embarrassed" that the Commission had taken more than four months to respond to a request from the subcommittee.[144] However, a staff member from that same subcommittee has stated that NRC responses frequently take from three to six months.[145]

In the spring of 1984, the Commission flatly refused to give Markey's subcommittee copies of transcripts of closed meetings on the Shoreham plant. (Commissioners Gilinsky and Asselstine opposed the decision.) This refusal was unprecedented in the full House Interior and Insular Affairs Committee's dealings with the NRC.[146] The Commission majority unilaterally decided that "no conceivable useful purpose could be served by breaching the confidentiality of that discussion" and, for the sake of providing "procedural fairness to all," urged that "the Subcommittee allow the administrative process to work, *free of close Congressional scrutiny at this time*."[147] Finally, under the threat of a subpoena, the Commission agreed to turn over the documents in an executive session of the full committee, meaning that all committee members would have to vote in order to make the documents public, which the committee later did.

One private Commission meeting revealed at least part of the reason why the agency has had rocky relations with Congress. The commissioners were discussing Markey's request for the transcript of a closed meeting on Shoreham. The head of the NRC's Office of Congressional Affairs, Carleton Kammerer, who was in charge of the NRC's responses to congressional inquiries, gave some advice to the Commission on the matter. He said that "what's at issue here is whether or not the giving of these documents would help the member of Congress make his case that the chairman has prejudged this issue."** Since the NRC's general counsel seemed to be saying that the documents could be withheld, Kam-

---

*See chapter 5, section on "Staff Investigation of Material False Statements at Diablo Canyon."

**See chapter 3, section on "Licensing Delay Scares of 1984."

merer stated, "*we ought to do what we should do to keep that document from this gentleman*."[148] At a hearing on Shoreham before Representative Morris Udall's subcommittee, Markey commented:

> Mr. Kammerer knew or should have known that the NRC has absolutely no right whatsoever to deny documents to congressional committees engaged in oversight investigations. Mr. Kammerer also appears to be advocating willful violation of the Atomic Energy Act because he believes the material in question would make the NRC Chairman look bad.
>
> To add insult to injury, Mr. Kammerer is in charge of the office that appears to be the main reason Congress isn't being kept fully and currently informed of a whole host of other issues.[149]

Which Plants Meet the Regulations?

In November 1979, in the wake of the TMI-2 accident, the House of Representatives passed an amendment to the NRC authorization bill offered by Representative Jonathan Bingham, which gave the NRC 120 days to report to Congress on the status of each operating plant's conformance to NRC regulations. The Bingham amendment required the NRC to identify the current safety requirements that were met by each operating plant, the generic safety issues for which technical solutions had been developed, and the licensed plants that had implemented the solutions.

The NRC commissioners, by a 3-2 vote, decided to oppose the amendment, which did not yet have final approval. The majority argued that to determine whether each plant complied with each NRC safety requirement would require 30 persons working six months. This expenditure was viewed as excessive, even though it represented only about 1 percent of NRC personnel resources.

Commissioners Gilinsky and Peter Bradford supported the amendment as "a necessary first step in developing a comprehensive program for the systematic evaluation of currently operating plants." They also expressed the view that the lack of readily available information on "the basic NRC safety requirements to which each operating reactor is subject . . . indicates a surprising disarray in the status of NRC knowledge of operating plants that should not be allowed to continue." Unfortunately, this disarray was allowed to continue.

The Commission majority opposing the amendment—Ahearne, Joseph Hendrie, and Richard Kennedy—offered an alternative: "a systematic safety evaluation of all currently operating plants." But they offered no schedule for completing the evaluation and only committed the NRC to a "progress report" prior to February 1981 and each year thereafter as part of its annual report to Congress.[150]

Section 110 of the NRC's fiscal year 1980 authorization bill passed by Congress contained a revised version of the Bingham amendment that required the NRC to identify regulations of "particular significance" to public safety; determine whether each plant met these regulations by following regulatory guides, staff technical positions, or "equivalent means"; identify the generic safety issues for which solutions had been developed; determine which of these solutions should be incorporated into the regulations; and provide a schedule for resolving the remaining generic issues.

In February 1981 the NRC staff was still developing its plan to implement Section 110, using what became known as the systematic evaluation program (SEP). The NRC staff sent the Commission an "outline of our revised plans" for reviewing the operating plants "in groups of 10 to 15 plants per year over the next eight years."[151] Thus, according to Representative Udall, the NRC managed to take a congressional request that "was originally intended to be carried out with a modest commitment of Commission resources and in accordance with a schedule lasting not more than 120 days"[152] and to turn it into "a multimillion dollar bureaucratic exercise that will not give final answers about the safety of today's operating plants until sometime in the 1990s."[153] The NRC has never complied with the Bingham amendment. To this day, it is impossible to determine the degree to which safety standards were applied to, or met by, any particular operating plant.

FIVE

SELF-REGULATION AND NUCLEAR CAMARADERIE

> We had a meeting with the NRC because of
> some apparent confusion . . . as to whose
> interests the agency was supposed to be
> representing. We were assured at that time
> by the General Counsel that the NRC was
> going to work with the Department of Jus-
> tice to ensure a more effective enforcement
> program. Aside from the question of any
> measurable improvement since then, estab-
> lishing an advisory committee on investiga-
> tive procedures at the request of agents for
> those who may be investigated and primar-
> ily if not entirely consisting of their attorneys,
> suggests that there may still be a serious
> identity problem despite previous assur-
> ances.
>
> Julian S. Greenspun
> Department of Justice, 1983

Much of the reason for the weak state of nuclear regulation is the fraternal relationship between the NRC and its licensees. The agency appears to view itself as an ally of the industry; it resists actions that place
the two in adversarial roles and has been willing to temper its criticisms
of the industry to the point where they lack effect.

Nuclear utilities are responsible for a great deal of self-regulation;
the NRC reviews only a small fraction of their work. The agency's failure
to provide an independent check on the utilities' design and construction
extends to its performance in licensing hearings. Agency rules make the
staff a party to the hearing, and the staff advocates issuance of the license.* The staff's advocacy role becomes primary, while its responsibility

---

*While advocating license issuance, the staff often recommends technical license con-
ditions.

to bring forth all relevant information, particularly that weighing against issuance of a license, recedes. The same tendency is seen in the NRC's investigative practices, which have frequently been inadequate or even obstructive.

The intimacy between the NRC and the industry it regulates goes beyond what the public and Congress have tolerated at other federal agencies. For example, a scandal erupted when a top-ranking Environmental Protection Agency (EPA) official turned over a copy of a draft report on dioxin contamination to Dow Chemical Company and changes desired by Dow were incorporated into the EPA's final report. The incident led to the official's resignation when it was made public by a congressional committee investigating the EPA.[1] Yet the NRC has released draft investigation, inspection, and evaluation reports to their subjects, and the Commission has consistently failed to take disciplinary action against the responsible officials.

The failure of the Commission to demand integrity and competence from both the utilities that operate nuclear reactors and the NRC staff, coupled with the degree to which the industry is trusted to police itself, has had costly and potentially dangerous results. In addition to the cases of Grand Gulf, Zimmer, and Diablo Canyon, which are described in this chapter, examples include Marble Hill, where construction and quality-assurance problems required major rework, leading to unaffordable delays and eventual cancellation of the reactors after $2.5 billion had been spent; Midland, where faulty construction work, improper ground preparation, inadequate rework, and quality-control problems put the plant twelve years behind schedule and resulted in its cancellation when it was 85 percent complete, at a loss of about $4 billion; Byron; Palo Verde; Comanche Peak; WPPSS-2; and South Texas, where hardware, construction, and quality-assurance problems led to costly delays but (thus far) a less severe fate.[2] At Three Mile Island, illegal manipulation of important safety tests by plant personnel was not made public until after the TMI-2 accident. Yet the NRC's subsequent failure to take firm action against the utility's management has done little to promote adherence to important safety requirements there or elsewhere.*

It is often difficult to determine whether regulatory problems stem from incompetence or from deficiencies in the agency's integrity. But whatever the cause, inadequate licensing review and investigations that

---

*See the section on "Falsification of Leak Rates at TMI-2" in this chapter.

conceal important facts neither serve the public interest nor brighten the future for nuclear power.

Inadequate Licensing Reviews

It is longstanding agency policy, dating back to the AEC, to conduct only an "audit review" of nuclear plant design and construction; indeed, the agency's resources are inadequate to do otherwise. The NRC staff reviews, at most, 40 percent of the design for one-of-a-kind plants. For plants whose designs are more familiar, the staff reviews significantly less. NRC inspectors physically review less than 1 percent of the completed plant to determine its compliance with NRC regulations.[3] The NRC's director of regulation, Harold Denton, explained that "using an audit process, it is simply not possible for the NRC to state, based on its own knowledge, that every rule and regulation has been met for every applicable action by the applicant."[4]

The degree to which the NRC relies on the industry to ensure safety is illustrated by Denton's conclusion that one reason regulation by audit is sufficient is that "the applicant for a license is obligated to assure compliance with applicable regulatory requirements."[5] The problem of limited independent review is compounded in licensing hearings because the NRC staff, as dictated by agency rules, is a party to the hearings and acts as an advocate for the utility.

Grand Gulf: Misuse of the Standardized Design Concept and the Perils of Self-Regulation

How could Grand Gulf—the largest nuclear plant in the world and the first intended to be built to a standardized design that was to be used in other plants for years—have been licensed to operate with hundreds of errors in its technical specifications and surveillance procedures?* Part of the answer lies in the willingness of the AEC and the NRC to cut corners in the licensing process to bring nuclear plants on line more quickly. The errors at Grand Gulf might have been found sooner had the NRC more thoroughly reviewed the utility's own safety analysis.

---

*Technical specifications are plant-specific requirements that reflect the NRC's assessment of which aspects of the plant must be closely controlled for the plant to operate safely. For example, the technical specifications identify the safety equipment that must be operable during reactor operation, list the required setpoints for initiation of automatic safety actions, and specify the type and frequency of surveillance tests of safety systems.

Mississippi Power & Light Company (MP&L) applied to build Grand Gulf Units 1 and 2 from a General Electric standardized design.* In the attempt to license the reactors promptly, however, events did not work out as planned. The way the AEC and the NRC handled the design's safety analysis report, or GESSAR, and the Grand Gulf plant that referenced the design illustrates how a potentially valuable concept can be derailed. It also stands as another example of delay induced, ironically, by the quest for speedy licensing.

To achieve its promised benefits, a standardized design should be in final form and approved before a construction permit that uses it is issued. Since the GESSAR was far from ready for a final design approval when some utilities anticipated using it, General Electric planned to obtain a preliminary design approval (PDA) that could be referenced in construction permit applications. But major portions of the preliminary design were still incomplete in 1974, when the AEC issued MP&L construction permits to build the two Grand Gulf reactors.

Before the AEC took this action, the staff's attorneys questioned the legality of issuing the permits based on the GESSAR design because in some instances the GESSAR contained no preliminary designs or even a basis for developing designs meeting the regulations.[6] (The NRC refused to release the memorandum in which this concern was expressed.[7]) The technical staff expressed a similar concern: "We are unable to even speculate on the nature of the proposed design and whether there is reasonable assurance that it can be implemented in accordance with the Commission's regulations."[8] The staff also raised concerns about potential problems with technical specifications at the operating license stage if construction permits were issued without an accurate description of the design features that were likely to be the subject of the final technical specifications.[9] Time showed this concern over the technical specifications to be justified.

Almost a year after the construction permits were issued, a high-level NRC official issued a memorandum that questioned whether many

---

*Standardized designs, once reviewed in detail and approved by the NRC, could be referenced by utilities in their construction permit applications without the need for a new review and approval. This process—if it were to utilize a complete, final design—would save time and help ensure that safety problems are resolved before rather than during construction or operation, when corrective actions are far more costly to implement. In France, plants operating since the mid–1970s have been standardized, contributing substantially to that country's ability to reduce construction time. Standardized designs also are used to a large degree in Japan. One widely recognized mistake of the American nuclear power industry is its failure to use standardized designs.

of MP&L's promises could be kept. The official worried that "construction permits were issued on the basis of design criteria for many systems, and commitments were obtained from the applicant without the benefit of a complete preliminary design to verify the practicality of these commitments."[10] As it turned out, the "commitments" were not kept. In some instances, the advanced state of construction at Grand Gulf made it economically infeasible to incorporate better engineering solutions to the GESSAR's safety problems once they became sufficiently well defined.

During construction, the utility proposed many changes in the preliminary safety analysis report on which issuance of the construction permits had been based. The NRC staff had stated that any changes in the " 'principal architectural and engineering criteria' which formed the basis for the issuance of the construction permit" would require (by the NRC's regulations) an amendment to the construction permit and a public hearing.[11] But in fact, many changes were made in these criteria without the issuing of construction permit amendments by the NRC and without public hearings. One change that had a far-reaching effect on the design was the reduction of the maximum earthquake acceleration that the plant had to be designed to withstand from 0.2 g to 0.15 g.[12]

When it came time to issue the operating license for Grand Gulf 1, the NRC showed that it had not learned that cutting corners can cost more time than it saves. Since the GESSAR technical specifications still had not been approved, the NRC began developing the Grand Gulf technical specifications from those applicable to an earlier-model GE boiling water reactor. MP&L compounded this error by hiring a consultant to assist in developing the technical specifications who had no experience in the operation of commercial BWRs.[13]

Shortly after the NRC granted Grand Gulf 1 a low-power operating license in June 1982, MP&L began to discover that the plant's technical specifications did not match either the actual equipment installed in the plant or the assumptions made in its safety evaluation. In other words, the fundamental premises of the safety evaluation were not ensured. While MP&L was required to propose technical specifications, the NRC is responsible for issuing the final version as part of the operating license. (At a meeting to discuss Grand Gulf in March 1984, the staff told the Commission that the NRC staff generally reviews "probably closer to five than fifty" percent of the technical specifications submitted by a utility.[14]) In October 1982, the NRC declared that Grand Gulf would not restart until the errors were identified. Yet the staff allowed the plant to resume low-power operation in September 1983.[15] Subsequently, many additional errors were found.

Representative Edward Markey, whose oversight subcommittee staff investigated the Grand Gulf problems, reported in July 1984 that in less than two years after Grand Gulf was granted an operating license, more than 1,000 errors were discovered in the technical specifications and related surveillance procedures, many of them with safety significance. The NRC had even approved license amendments referring to nonexistent equipment. Markey charged that "the NRC lifted its ban on operations prior to adequate assurance that this part of the license had been fully reviewed and corrected."[16]

In July 1984, the Commission voted to grant Grand Gulf a full-power license. It acted in part on an assertion by the regulation director, Denton, that "I think we are now confident this set of tech specs does reflect the application, does reflect the plant, does reflect the safety analyses that are contained in the application."[17] The staff also pledged that MP&L would be required to formally certify the accuracy of the revised technical specifications, which the utility did on August 5. But two weeks after the NRC staff vouched for the technical specifications, the utility informed the NRC that ten circuit breakers, built into the plant and assumed in the utility's safety analysis, were not listed in the technical specifications as they should have been.[18]

The many errors found at Grand Gulf might have been discovered earlier if the NRC had performed a full review of the utility's safety analysis to confirm the accuracy of the necessary technical specifications. (Such a review was urged by Commissioner Victor Gilinsky, the Union of Concerned Scientists, and others but was rejected by the Commission majority.) Furthermore, NRC regulation 10 CFR 50.36(b) requires that technical specifications be derived from the utility's safety analysis, which clearly had not been done at Grand Gulf.

The lack of accurate technical specifications was not the only problem encountered in Grand Gulf's troubled history. According to Markey, operator training records were apparently falsified (a matter turned over to the Department of Justice); at least five reactor operators were removed from active duty because they were found to be unqualified; 238 "mishaps" that required immediate notification of the NRC's Emergency Operations Center occurred in less than two years, a rate the NRC called abnormally high; the plant management received low ratings from the NRC for two consecutive years; and significant design errors were discovered in several important safety systems.[19] One would expect that the licensing of such a plant would proceed with the utmost caution. Instead, the NRC, claiming that Grand Gulf's management had improved, granted the plant a full-power license while problems were still being discovered.

Licensing Hearings: The NRC Staff as the Applicant's Advocate

Independent and thorough scrutiny of nuclear plant design and construction is frustrated by the NRC staff's performance as a party in licensing hearings. As described in chapter 3, the staff has resolved whatever major differences it may have had with the utility by the time the hearing begins. From then on, with rare exceptions, the staff becomes an advocate for issuing the license. Having completed its review, the staff is in the position of defending its prior determinations and is reluctant to consider that the safety issues raised by intervenors may have merit and might have been overlooked during the staff's safety review. The determination of the staff as advocate to have its view prevail leads sometimes to the presentation of far-fetched arguments and incomplete or misleading information to the licensing boards.

One licensing board chairman, Herbert Grossman, expressed his frustration with the staff's lack of independence. In dissenting from a decision concerning the proposed restart of a test reactor at the Vallecitos Nuclear Center in California, Grossman wrote: "On the record before us, it is difficult to distinguish between [NRC] Staff's presentation and that of a typical private litigant, whose counsel might be expected to present only evidence favorable to its position and to caution its witnesses not to volunteer unfavorable information or opinion."[20] Two other cases—Byron and UCLA—demonstrate the effects of the NRC staff's role as advocate for the license applicant in hearings.

Byron. The nuclear industry shuddered when on January 13, 1984, an NRC licensing board denied—for the first time ever—an operating license to a completed nuclear power plant. The surprise was compounded because the plant, Byron 1 in Illinois, was owned by the Commonwealth Edison Company, which owned more reactors than any other utility. The denial involved not only widespread quality-assurance failures by the company itself (specifically, loose supervision over the work performed by certain key contractors) but also a failure of the NRC staff to communicate material information to the licensing board.

In denying the license, the board focused on the NRC staff's failure to inform it until the hearings were virtually over—and then essentially as an incidental matter—that it had ordered Commonwealth Edison to undertake a comprehensive reinspection program. The board found that the results of this reinspection program were central to determining whether Commonwealth Edison's overall quality-assurance peformance was adequate—the key remaining issue in the licensing proceeding:

> The relevance and importance of the reinspection program at Byron to the licensing review of the Byron plant, and to the quality assurance litigation is, and has been, obvious. Yet when the Region III staff first prepared its direct testimony on the quality assurance contention . . . it made no mention whatever of a reinspection program.[21]

The board noted that although the staff depended in part on the reinspection program to resolve charges made by witnesses during the licensing proceeding, "the Staff was not able to provide assurances to the Board that the reinspection program was adequate." The board also cited staff testimony stating that the staff would not make a final assessment of the adequacy of the reinspection program until up to three months after Commonwealth Edison submitted its results and evaluation. The board added that "the Staff's original presentation, totally ignoring the recertification and reinspection program, has never been explained by the Staff nor understood by the Board."[22]

The board attributed the staff's "slighting of the issue of the reinspection program" to "the Staff's misunderstanding of the respective roles the Staff and the licensing boards play in the licensing process. The Staff appears to think the Board can delegate to it the responsibility of deciding the essence of the issues raised by the contention on quality assurance."[23] The board disagreed with this view, because accepting it would have essentially rendered meaningless the public's right to an independent adjudicatory hearing.[24]

Commissioner Gilinsky later told a congressional subcommittee that in this case "the staff, which was seeking approval of the license, tried to circumvent the Licensing Board review of the plant's quality-assurance plan and failed to inform the Board promptly of some of the problems at the plant. As it turned out, this approach backfired."[25] After the license was denied, the staff met with Commonwealth Edison to discuss, in Gilinsky's words, "how best to extract the license from the Boards."[26] After further board hearings, the license was granted.

The UCLA Research Reactor. The NRC staff's advocacy role in hearings led it to make questionable statements to the licensing board in the case of the research reactor at the University of California at Los Angeles (UCLA).

In March 1981, an intervenor (Committee to Bridge the Gap) raised the issue of whether the reactor had adequate protection against sabotage—an issue that took on added significance because the 1984 Olympics, potentially an attractive stage for terrorists, were to be held nearby. For approximately three years, UCLA and the NRC staff told the licensing board that NRC regulations did not require such protection for nonpower reactors. However, the licensing board ruled that "10 CFR 73.40(a) requires that UCLA take some measures to protect against potential sabotage."[27] The issue remaining to be litigated was whether that protection was adequate.

In February 1984, the licensing board reviewed the reactor's physical security plan and its security inspection reports for 1975–79 and 1982. The board's reaction:

> We were astounded to read in the first sentence of the first paragraph of the test of UCLA's physical security plan that it was indeed the purpose of the plan to provide "for the protection of the reactor, protection of the staff and the general public against radiological sabotage. . . ."

> We were even more astounded to find that every Part 73 security inspection report furnished by UCLA indicates that Staff did, in fact, examine UCLA's activities related to physical protection against sabotage "in accordance with applicable requirements of Title 10, Code of Federal Regulations, Part 73."[28]

Moreover, on November 9, 1983, the NRC staff had amended the license to require UCLA to "maintain and fully implement *all* provisions" of its physical security plan.[29] About a month after the licensing board informed the parties of these findings, the staff's attorney told the board that an NRC manual instructs inspectors to inspect research reactors for protection against radiological sabotage and that such an inspection had occurred the previous November at another university's reactor.[30]

The board was particularly critical of misleading statements made by NRC technical staff members in affidavits supporting the staff's April 1981 request that the issue be dismissed (based on the premise that protection against sabotage was not required). The affidavit of James Miller, a branch chief in the NRC's Division of Licensing, asserted that he had "verified"[31] that the reactor met an NRC standard that exempts research reactors from certain protection requirements if the irradiated fuel is so highly radioactive that it is "self-protecting," that is, it cannot be handled without the risk of severe injury. To qualify for this exemption, the fuel must not be readily separable from other radioactive material and must have a radiation dose rate in excess of 100 rems per hour at a distance of three feet.[32] The staff's assertion that the exemption applied was critical to the reactor's licensing; without the exemption, UCLA would have had to comply with protection standards that it did not meet.[33]

In fact, two letters from UCLA to the staff and a staff paper to the Commission stated that UCLA could not meet the exemption conditions. The staff and UCLA argued that these statements were made before some fuel was removed and that new calculations showed the exemption to be met. However, the new calculations were based on the dose from the core as a whole, not from each individual fuel bundle. The licensing board recognized that a number of statements from NRC staff members, as well as the language in the NRC regulation, indicated that each fuel bundle must meet the standard.[34]

The board made its displeasure clear: "[W]hen an affidavit stating a conclusion is furnished, that affidavit must state precisely what the conclusion is and on what basis it is founded. Mr. Miller's affidavit . . . did neither."[35] The board found that Miller's affidavit did not make clear that he had determined UCLA's fuel to be self-protecting based on the dose rate of the entire core, nor did it explain why he adopted this approach rather than computing the dose rate for each individual bundle. The board concluded that the "Staff's failure to furnish this sort of information in the first instance certainly results in delay and a waste of time at minimum and, at most, a loss of confidence in the licensing proceeding and a board decision which is not well founded."[36]

Another affidavit discussed by the licensing board was submitted by Donald Carlson, a plant protection analyst in the NRC's Division of Safeguards. Carlson's affidavit stated that "[t]here are no explicit NRC regulations for the protection of non-power reactors against radiological sabotage. . . ."[37] The licensing board concluded that this statement was "plainly wrong" and that, while the Division of Safeguards may have felt that protection against sabotage was unnecessary, such a position was contrary to the practices of the agency's Office of Inspection and Enforcement and was not accepted by the Commission.[38]

The board pointed out that the Division of Safeguards now claimed that a regulation passed in 1979, 10 CFR 73.67, superseded the regulation requiring protection against sabotage, 10 CFR 73.40(a). Yet Carlson himself stated in a meeting that section 73.40(a) still required protection against sabotage, and no one expressed disagreement with this position. In addition, both UCLA's draft security plan and the cover letter transmitting the plan mentioned this requirement. While concluding that the affidavit did not contain a lie, the board acknowledged that the division's position amounted to a repeal of section 73.40(a) and pointed out that "[s]uch a repeal cannot properly be made by Staff acting unilaterally."[39] The board reminded the staff that "[j]ust as anyone else, Staff is bound by the rules. Until such times as they are amended, Staff must follow the rules."[40]

The licensing board also criticized Carlson for failing to inform staff counsel and, through it, the board, that the security plan contained provisions addressing protection against sabotage:

> . . . Staff has an ironclad obligation to bring relevant and material information to the attention of boards. Mr. Carlson's failure to advise Staff counsel of the provisions in the UCLA Security Plan of the very sort we had held to be required presents a situation that cannot be tolerated in NRC adjudication. Staff, as the keeper of the public trust, must be particularly sensitive to this obligation.[41]

Two other NRC staff affidavits also contained questionable statements. After the licensing board's February order asking the staff to defend its position on sabotage in light of the provisions in the plan, Leroy Norderhaug, chief of an NRC Region V safeguards branch, and Matthew Schuster, chief of a security section, both told the board that after 1979 research reactors were no longer inspected for sabotage. This statement was inconsistent with a 1982 inspection report for the UCLA reactor, an inspection in 1983 of another research reactor, and the NRC Inspection Manual, Chapter 2545. The board wrote:

> Affidavits should only be executed after the affiant has carefully ascertained the facts sworn to. Obvious, unexplained inconsistencies between an affidavit and established Staff procedures . . . cannot be tolerated. Boards must . . . be able to rely absolutely on Staff's representation of factual matters. There is simply too much at stake in our adjudications to permit mistakes of fact, particularly by the NRC Staff. Staff affidavits which are ambiguous or incorrect force boards to engage in time-wasting inquiries to determine the facts or risk rendering a decision based on ambiguous or incorrect information.[42]

In the end, the board noted that it lacked the authority to discipline the NRC's technical staff but felt it necessary to bring the "improper practices" to the Commission's attention for whatever action the Commission deemed appropriate.* The board added that insufficient evi-

---

*At this writing, the Commission had not responded to the board's findings. A report on this case by the NRC's Office of Inspector and Auditor was before the Commission but was not available to the public.

dence was available to "conclusively show misconduct."[43] Nevertheless, the board did point out that the staff's desire to prevail had caused it to abandon its responsibilities to the public:

> . . . while we cannot know specifically what may have led to the concerns we have identified above, we fear that a contributing cause may have been Staff's embroilment in this proceeding. It is understandably hard to remain detached when one's positions are attacked. However, Staff's obligation is to the public interest, and its member should take care that their actions are directed toward that end rather than toward besting an adversary.[44]

Questionable Investigative Practices

The close relationship between the NRC and the industry it regulates is nowhere more apparent than in the agency's investigative practices. The NRC has conducted investigations that are half-hearted, obfuscatory, or less than independent. The agency's investigative practices have included premature releases of draft reports to the reports' subjects, breaching whistleblowers' confidentiality, thwarting potential probes by the U.S. Department of Justice, and, more generally, failing to conduct thorough and unbiased investigations.

The Office of Inspection and Enforcement (IE) inspects nuclear plant sites to verify compliance with NRC regulations. The Office of Investigations (OI), created in April 1982, is responsible for investigations of possible licensee misconduct and for recommending referral of potential criminal cases to the Department of Justice. While NRC investigative practices improved with the creation of the OI, evidence shows that this office does not have the independence and resources needed to perform the job effectively.

Serious deficiencies in NRC investigations have been uncovered by congressional committees, members of the public, and the NRC's Office of Inspector and Auditor (OIA). The OIA is the NRC's limited version of an inspector general, auditing NRC programs and investigating questionable practices of the NRC staff. Unfortunately, according to one congressional subcommittee, "the watchdog's watchdog has a short leash,"[45] and some OIA internal audits have been restrained by the Commission. Perhaps worse, the OIA has shortened its own leash. Although it is charged with investigating questionable staff actions, it has mimicked some of the staff's poor practices in carrying out its investigations.

Sharing Draft Investigation and Inspection Reports

Investigation of the TMI-2 Accident. After the TMI-2 accident, the NRC staff conducted an investigation into the accident's causes and ramifications. In 1979, the staff gave a draft copy of its report to the plant's operator, General Public Utilities (GPU). During a 1982 lawsuit between GPU and the reactor's manufacturer, Babcock & Wilcox, a draft of the report with GPU's comments, obtained by Babcock & Wilcox, indicated that the NRC staff, at GPU's request, had deleted information about safety problems at the plant from the final report.

One excised discussion concerned the potential for failure of the emergency feedwater system, the only system able to cool the reactor following an accident like that at TMI. The deleted section described a January 1979 event at TMI-2 that filled the area containing all three emergency feedwater pumps with steam. The NRC's draft report recommended that "the steam release should be evaluated as a potential common mode failure mechanism for the emergency feed system. This item is considered unresolved."[46] In other words, the problem could disable the emergency feedwater system and the NRC did not yet have a solution.

Another deleted section, called Concerns, had revealed that information about possible problems with the pilot-operated relief valve (PORV), the component that stuck open during the TMI-2 accident and allowed hundreds of gallons per minute of cooling water to escape, had been given to the utility in 1976. The draft report had referred to a letter from the PORV manufacturer, Dresser Industries, that described conditions under which the valve might fail:

> In particular, the attachment to this letter describes some features of the electromatic relief valve that raise questions concerning the suitability of this valve for low-pressure [reactor coolant system] protection, the second use of this valve on the pressurizer.[47]

The director of the NRC's Office of Inspection and Enforcement, Victor Stello, Jr., whose office prepared the report, later said the removal of this section was justified because the problem with the valve was "generic"—shared by other plants using valves of a similar design.[48]

The Hayward Tyler Investigation. The Hayward Tyler Pump Company case was one of a number of questionable NRC investigations examined by the staff of a House Interior and Insular Affairs subcommittee. The case began in December 1981, when five former employees of Hayward

Tyler sent the chairman of the subcommittee, Representative Markey, allegations of serious manufacturing and quality-control deficiencies in cooling pumps that Hayward Tyler was marketing to commercial nuclear power plants. The ex-employees asked Markey to request an NRC investigation,[49] claiming that an earlier NRC investigation was inadequate because the NRC looked only at the company's paperwork, which had been tampered with, and not at the hardware itself.[50] Markey did so, and an investigation was initiated through the NRC's Region IV office.*

NRC investigators visited the Hayward Tyler plant in January 1982 and reviewed the company's documents. At the end of that month, Deputy Regional Administrator Karl Seyfrit ordered the chief investigator, Richard Herr, to terminate the investigation, despite Herr's objections that it was incomplete.[51] Region IV Administrator John Collins had met earlier that week with attorneys representing Hayward Tyler and reportedly told them that the investigation would be terminated that week.[52] One of these attorneys was Marcus Rowden, a former NRC chairman. Rowden retained as a technical consultant the former director of the NRC's IE office, Ernst Volgenau. Volgenau retained as his consultant Dudley Thompson, a former IE director of enforcement and investigations.[53]

On February 12, 1982, the NRC Region IV staff met with Hayward Tyler representatives at the company's request. The staff released to the Hayward Tyler attorneys copies of the draft inspection and investigation reports and the transmittal letter and notice of nonconformance listing violations for which the NRC planned to cite the company. The decision to release these documents was made by Collins and Seyfrit. Twelve days later, Seyfrit gave a second draft inspection report, transmittal letter, and notice of nonconformance to Rowden, who suggested that certain changes be made in the letter of transmittal. Some of these changes, which softened the letter, were made by the NRC staff.[54] Later, Collins conceded:

> To the best of my knowledge, it has not been NRC practice to provide copies of draft investigation or inspection reports to the parties being investigated for review before issuance. In this case, however, I felt that review would enable the vendor to immediately take corrective action on the findings and provide a mechanism for immediate release of the reports to the public once they had been finalized. It should be recognized that although this action may have been precedent setting, there is no written policy or guidance from Headquarters covering these types of actions.[55]

---

*Region IV was responsible for conducting inspections and investigations of vendors of nuclear plant equipment.

However, three investigators from the OIA later cited several NRC regulations and procedures that the releases had violated.[56]

At a meeting six days after Collins had released the first set of draft material to Hayward Tyler, the NRC's executive director for operations, William Dircks, gave Collins "unequivocal direction" to make sure that neither the draft investigation nor the draft inspection reports would be released. Collins apparently did not mention then that he had already released the drafts to the company.[57] More troubling is the fact that six days after Dircks issued this directive, the NRC staff released the *second* set of draft materials to the company. Moreover, one week after issuing this warning, Dircks and the NRC's executive legal director, Guy Cunningham, met with Rowden and discussed problems Rowden had with the transmittal letter for the report. According to NRC internal investigators, these high-level officials were "confronted with information which unquestionably established that Rowden . . . had received copies of draft Region IV documents" but "failed to recognize that there were violations of either NRC regulations or procedures."[58]

In reviewing this incident, the OIA found that while the release should not have occurred, the NRC lacked "comprehensive policy and procedures with regard to field investigations."[59] Chairman Nunzio Palladino, in summarizing the OIA's findings, told a congressional hearing:

> the Inspector and Auditor found that, notwithstanding whether or not we have specific regulations or guidance addressing this point, common sense should have dictated that the draft reports should not have been released by the Region. He also found that NRC contacts with company representatives were at less than the proper arms length distance, and they were not adequately documented.[60]

Palladino added that the "arms length" situation referred primarily to Dircks's support of changes in the letter transmitting the inspection report, which occurred after Dircks was urged to change the letter by Rowden at an earlier meeting.[61] Palladino also said that OIA Director James Cummings had found that "as a result of company contacts, NRC's proposed transmittal letter had been softened," but that the NRC made the changes "in good faith," believing that they were "on point and valid."[62] The OIA also concluded that the NRC should have kept the company at a distance "not only because of the competing interest that exists during an investigation but particularly because of the fact that NRC officials were dealing with former associates who were also former senior officials of the NRC."[63]

Yet, in spite of these conclusions, Cummings recommended that no disciplinary action be taken. In fact, three OIA investigators assigned to the Hayward Tyler case had written an internal memo to Cummings advising that their investigation had revealed that "a number of NRC personnel made decisions which were clearly contrary to existing regulations and manual guidelines. . . . [I]t is clear that the facts in this case do warrant *some* disciplinary action."[64] Cummings, however, rejected his staff's recommendations, responding that *"[s]tupidity, lack of common sense and/or incompetency does not in my mind necessarily warrant disciplinary action. . . . If disciplinary action resulted from every screw up in NRC, few of us would be around."*[65] Markey called for a "housecleaning" at NRC, referring mainly to Region IV and to Collins and Seyfrit, in particular.[66] No disciplinary action was taken against anyone for the Hayward Tyler incident.[67]

Because of the improper release of the draft report, the case was reinvestigated in December 1982, this time by the NRC's newly formed Office of Investigations. The OI report substantiated several of the original allegations but made no finding on other allegations because the information necessary to assess their validity was no longer available.[68]

New Releases of Draft Reports. In the aftermath of the Hayward Tyler case, Representative Markey told the NRC that the agency's release of the draft material had further undermined the public's confidence that the NRC was working foremost for public safety. Chairman Palladino assured the subcommittee that efforts would be made to prevent a recurrence.[69] Yet later events revealed that the Hayward Tyler release was not to be the last.

One case involved a joint investigation/inspection of quality-assurance violations at the Washington Public Power Supply System (WPPSS). In April 1982, a Region V inspector, Paul Narbut, released to WPPSS a draft inspection report (considered in part an investigation[70]) and a draft notice of violation. Narbut apparently had not yet received a March 24, 1982, memo from Dircks, warning the staff that "under no circumstances will *draft* reports of investigation[s] be reviewed with or given to licensees or their agents without my express permission."[71] The Narbut release occurred a day or two before Region V Administrator Robert Engelken distributed the Dircks memo (on April 9, almost two weeks after Dircks had issued it) along with his own instructions that extended the prohibition to inspection reports.[72] On July 30, 1982—almost four months after the report's release to WPPSS—Dircks issued a memorandum to NRC regional administrators directing that no draft *inspection* reports be released without his permission.[73]

Upon learning of the incident involving WPPSS, Markey requested from the NRC a list of all other draft reports that had been released to utilities.[74] The NRC found that only one draft *investigation* report other than in the Hayward Tyler case had been released since 1981, but more than twenty draft *inspection* reports had been released.[75] Some of these releases occurred *after* Dircks distributed his July 1982 memorandum instructing that draft inspection reports not be released without his express permission. In responding to Markey's request, however, Chairman Palladino neglected to cite the twenty reports, writing only that "a number of draft inspection reports have been released."[76] In fact, the only report specifically mentioned was one Markey had already been informed of.[77] This omission appears to have been intentional. One internal NRC memorandum listing the twenty inspection reports contained a handwritten note that "although this is quite specific regarding facilities, the response to Markey probably doesn't need the specificity."[78]

One of these twenty premature releases occurred at San Onofre. There, in March 1982, the NRC staff released an entire draft of an inspection report concerning safeguard deficiencies to the plant's operator, Southern California Edison. That occurred with the approval of the regional enforcement director and the Division of Safeguards director, at a time when the NRC was actively considering enforcement action, including a $20,000 fine.[79] This action was directly contrary to NRC procedures as specified in the Inspection and Enforcement Manual:

> Advance copies of inspection/investigation reports provided to headquarters in support of escalated enforcement action should not be sent to licensee/vendors for review in accordance with this chapter until the documents initiating the enforcement action have been signed and issued.[80]

As a result of the premature release, the utility responded in a manner that was apparently in part responsible for the NRC's subsequent decision not to take any enforcement action at all. The staff of the congressional subcommittee that reviewed this case concluded that while the safety problems at the plant may have been resolved, the way the NRC handled this case violated the agency's rules and thwarted an important purpose of enforcement actions and civil penalties: to bring problems to the attention of the public and focus scrutiny on utility management.[81]

As in the Hayward Tyler case, the director of the OIA, Cummings, overrode his principal investigator's recommendation that disciplinary action be considered by the Commission because of an "obvious violation of an unambiguous procedure by senior regional managers."[82] Not only did Cummings omit the investigator's recommendation from his memorandum to the Commission (which he had the authority to do) but, according to the investigator, he also "improperly state[d] that there was unanimity" on his recommendation that no sanctions were warranted.[83] In submitting the investigation report to the Commission, Cummings had written, "In *our* opinion, no administrative sanctions are warranted,"[84] and put the investigator's name on the concurrence line.[85]

As late as 1983, despite congressional criticism and repeated guidance from upper management, the NRC staff still had not stopped giving draft reports to licensees. For example, details of a draft report, in the form of inspection notes, were released in April 1983 to the California utility operating Rancho Seco. Region V Administrator John Martin found that while "the entire staff was well aware of the prohibition on release of draft reports to persons outside the agency," many felt that releasing inspection notes was permissible. Martin believed that a release such as that at Rancho Seco violated the "spirit and principle" of the prohibitions of the executive director for operations on releasing draft reports.[86] Thomas Rehm, of the operations office, confirmed that while these notes were not exactly a draft report, they "probably [go] beyond what should be released in writing."[87]

Also in 1983, the NRC released to the respective utilities draft reports of studies of quality assurance at three construction projects. The Diablo Canyon draft was given to Pacific Gas and Electric Company (PG&E) even before it was given to Congress, for whom the studies were being prepared. Yet when the draft was later given to one congressional staff member, he was instructed to keep it confidential.[88]

In a letter of September 2, 1983, the NRC's director of licensing, Darrell Eisenhut, told PG&E's executive vice-president that the utility's comments were to be considered "in connection with [NRC's] finalizing of the Case Study."[89] When the propriety of providing a draft to the company was questioned by Representative Markey, the NRC claimed that "it would have been reasonable" to solicit PG&E's comments "to correct errors of fact and misinterpretation," but that the NRC had decided not to do so "because of concern over the appearance of impropriety. . . ."[90] Nonetheless, the NRC did not deny that PG&E's comments were received. Instead, it vigorously maintained that the people writing the report were not aware of PG&E's comments. [91] The NRC did not explain why comments were solicited if they were not provided to the persons drafting the report.

While the draft had criticized PG&E's quality-assurance program "in blunt, uncompromising terms," the final report included only a few of the draft's strongly negative conclusions and the language was toned down, according to Markey.[92] Whether or not PG&E's comments were actually considered in writing or in an NRC management review of the final study, the NRC's action in soliciting the utility's comments on the draft created a strong appearance of impropriety.

Moreover, while the draft was directly relevant to ongoing Diablo Canyon licensing proceedings, the NRC did not give it to the other parties in the proceeding or to the licensing board until about a month after it was given to the utility. The intervenors only obtained the draft after learning of its existence and requesting if from the utility and the NRC under discovery procedures.[93]

In responding to concerns expressed by Markey over this incident, Chairman Palladino agreed that it was the NRC's responsibility to be alert to new information that could affect proceedings and to transmit this information promptly to the licensing board, and he acknowledged that "NRC procedures were not followed in this regard."[94] Once again, however, the Commission took no disciplinary action.

Breaching Confidentiality

Another troublesome element of NRC investigations has been the release of the identities of whistleblowers—nuclear plant workers who bring problems in confidence to the NRC's attention. The NRC breached the confidentiality of Charles Atchison, a former quality-control inspector for Brown & Root, which at the time was the major construction contractor for the Comanche Peak plant in Texas. On April 12, 1982, Brown & Root fired Atchison. Just one day later, the NRC's senior resident inspector, Robert Taylor, divulged to the utility that Atchison was the source of allegations made in 1980. Asked why he had breached confidentiality in this way, Taylor responded that the NRC's Region IV office and probably the other regional offices had received "unwritten guidance" that the agency would shield confidential informants only as long as they were employed at the plant site. Taylor called this a "common sense policy" because the informant would no longer be subject to dismissal.[95] Such a "policy" ignores the fact that the former employee might attempt to find work elsewhere in the nuclear industry.

In July 1984, the confidentiality of another former Comanche Peak worker was breached when the NRC sent his deposition to the NRC's

Public Document Room with his name on it. His name was available to the public for six days before a citizen group (Citizens Association for Sound Energy) participating in the case discovered it and had it removed. While this breach may have resulted from an error, the effect was a potentially damaging violation of a protective order guaranteeing the employee's confidentiality.[96]

At Diablo Canyon, the NRC divulged the identity of a confidential whistleblower by referring in a public report to "the alleger" as "a Pullman Quality Control Inspector" who wrote "Deficient Condition Notice No. 1604-006."[97] The identity of the author of this report was readily available to PG&E.

The NRC's practice of turning over unpublished evidence to the respective utilities, including whistleblowers' affidavits and other documents, also has led to the identification of confidential whistleblowers. Although the whistleblowers' names are whited out, the staff affidavits contain highly detailed information that utilities can match with issues raised by individuals on site, making anonymity highly unlikely. PG&E was apparently able to identify at least three confidential whistleblowers in this way, referring to them by name in responding to their allegations.[98]

Frustration of Department of Justice Investigations

In two cases involving nonreactor licensees, Nuclear Pharmacy Incorporated and Stepan Chemical Company, the NRC was accused by the Department of Justice of bungling potential criminal cases. According to a congressional subcommittee staff, Nuclear Pharmacy had repackaged chemical-grade, "not for human use" Xenon-133 and sold it to hospitals for human use, and Stepan Chemical had failed to inform the NRC of an allegedly radioactive waste burial site underneath a building in which the secret ingredient for Coca-Cola apparently is stored.[99]

In the Nuclear Pharmacy case, after referring the matter to the Justice Department for possible criminal prosecution, NRC personnel, including IE Director Victor Stello, Jr., met with Nuclear Pharmacy and, according to the Justice Department, "disclosed Government theories and evidence." The Justice Department found that "these disclosures have the potential of materially impairing our ability to obtain untailored grand jury testimony or interview statements."[100]

In the Stepan Chemical case, in spite of a memo from the NRC's general counsel that asked "whether the matter should be referred to the [Justice Department] for potential criminal action," Stello initiated an IE investigation. According to the Justice Department, by the time the department was alerted, IE had "interviewed and debriefed the suspected individuals, thus increasing the likelihood of viable alibi defenses."[101]

In both cases, the Justice Department found that the NRC had so reduced the possibility of bringing charges that the cases were no longer worth pursuing. Furthermore, the disclosures to Nuclear Pharmacy apparently occurred after the NRC had promised the Justice Department that it would not touch the case until the department had completed its criminal investigation. In a stinging letter to the NRC, Lawrence Lippe, chief of the General Litigation and Legal Advice Section of the Justice Department's Criminal Division, observed that the agency might have forgotten whom it existed to protect:

> gratuitous vacillations of positions, gratuitous and apologetic concessions and admissions, and gratuitous disclosures of evidence and theories prior to the completion of an investigation, could deservedly or undeservedly suggest a basic confusion as to who the respective clients are. The clients, of course, are the public—the millions of persons who cannot afford to retain lawyers and law firms to represent them before the NRC and who have no other persons to protect their interests. Furthermore, if any benefits of doubt are to be made, they should be made on the side of public safety. . . .[102]

The Justice Department was also at odds with the NRC over the agency's prolonged suspension of its investigation into falsification of leak-rate tests at Three Mile Island-2* and expressed frustration at the NRC for stalling its referral to the Justice Department of potential criminal violations at Zimmer.** In March 1983, the deputy chief for litigation of the Justice Department's Criminal Division, Julian Greenspun, again criticized the NRC's investigative policies. Greenspun pointed to the past "reluctance, if not resistance from certain NRC quarters insofar as developing criminal cases" and noted that he had been encouraged by the NRC's expressed interest in improving its enforcement program and by the creation of the Office of Investigations in mid-1982.[103]

After these hints of improvement, however, Greenspun had learned of a new advisory panel appointed by the Commission to propose guidelines for NRC investigations. This panel was formed at the request of and included attorneys who represented nuclear utilities or had previ-

---

*See the section on "Falsification of Leak Rates at TMI-2" in this chapter.
**See the section on "Quality-Assurance Breakdown at Zimmer" in this chapter.

ously done so. Greenspun wrote the NRC that the establishment of this panel:

> detracts not only from verbal statements about revitali⁊ing (or establishing) an effective enforcement program, but also from the wisdom of solely depending upon the NRC to investigate possible violations within a district (as opposed to using other agencies assisted by nuclear experts retained as investigative consultants, as is done by the Department in other kinds of investigations).[104]

Greenspun said he was "astonished" to read a letter from a utility lawyer requesting the establishment of the panel and asking the Commission to prevent NRC investigators from interviewing company employees without first notifying company management. The letter also requested the Commission to establish other procedures that would, according to Greenspun, "inject company management and attorneys" between NRC investigators and company employees. Greenspun reminded the NRC of the difference between its role and the roles of utility management and attorneys:

> as its name implies . . . the Nuclear Regulatory Commission was created by Congress to represent and protect the public interest. . . . While the power companies, which are not charitable institutions, have their lawyers to represent their interests, the public has no one to represent theirs in this area of such vital concern to them, except the NRC, the Department of Justice and the United States Attorneys Offices. In representing the public, the NRC's job in conducting inspections and investigations is to find out the truth, vigorously and within the boundaries of the law, in order to find out if the laws and regulations are being complied with and determine if they are not. The NRC has nothing to be ashamed of, or apologize for, for doing this—only if they do not.[105]

Greenspun concluded that the NRC action "suggests that there may still be a serious identity problem despite previous assurances."[106]

Inadequate Investigations

Falsification of Leak Rates at TMI-2. In November 1983, the General Public Utilities subsidiary, Metropolitan Edison (the predecessor of General Public Utilities Nuclear as the Three Mile Island license holder), was indicted by a federal grand jury on eleven criminal counts. In February 1984, the utility pleaded guilty to one count and no contest to six others. The charges involved the falsification of leak-rate tests (and related records) that are used to determine the rate of leakage from the reactor's primary cooling system. For months before the accident at TMI-

2, plant operators had been routinely discarding test results that showed the leak rate was greater than allowed by the plant's license. They had also been improperly manipulating tests to make them appear to show leak rates within acceptable limits.[107] If the utility had adhered to the terms of its license, TMI-2 would have had to shut down. There is much evidence to indicate that during the accident that began on March 28, 1979, plant operators disregarded some indications of high temperatures, attributing them in part to the leakage that they had become familiar with.[108] Operators thus did not learn for more than two hours that the pilot-operated relief valve was stuck open, allowing hundreds of gallons per minute of cooling water to escape.

In October 1978, five months before the TMI-2 accident, an NRC inspector had discovered that TMI-2 employees had not been reporting "bad" leak-rate tests. The inspector told top plant personnel that he found this "shocking" and a "fundamental misinterpretation of the safety requirement." The company assured the inspector that the practice would be changed.[109] No enforcement action was taken against Metropolitan Edison. According to U.S. Attorney David Queen, the situation at TMI-2 grew progressively worse instead of being corrected and, after January 1979 until the time of the accident, virtually all leak-rate calculations had to be manipulated to get acceptable results.[110]

In May 1979, a former operator at TMI-2, Harold Hartman, told the NRC's Office of Inspection and Enforcement that operators at TMI-2 had routinely "fudged" leak-rate tests prior to the accident.[111] The IE office formally interviewed Hartman but undertook no further investigation. No mention of these allegations was made in the extensive IE investigation report of the accident, NUREG-0600. IE passed on its files, including the Hartman interview, to the NRC's Rogovin group, which was investigating the accident. Harold Ornstein, an NRC staff member on detail to the group, investigated further, including interviewing Hartman. Ornstein drafted a section for the Rogovin report that substantially confirmed the charges, but it was not included in the final report.[112]

The NRC appears to have taken no further action on the Hartman allegations until March 1980, when television station WOR in New York broke the story. The NRC then began to conduct interviews. Soon after, the NRC referred the matter to the Department of Justice and in May 1980 suspended its own investigation. The NRC later maintained that the Justice Department had requested the NRC to suspend its investigation and that the NRC had been unaware that it could resume its investigation until May 1983. The Justice Department countered that at most it had requested that the NRC not interview certain people for a limited period. Moreover, Assistant Attorney General Lowell Jensen, head of the Criminal Division, stated that the Justice Department specifically told the NRC in October 1981 that "the NRC should feel free to proceed" with its interviews, with the exception of three specific individuals.[113] The NRC did not resume its investigation until late May 1983, seventeen months later. In January 1985, the Justice Department wrote Representative Morris Udall that it believed that it "was misled by the NRC . . . regarding the NRC's knowledge of our conveyance in the fall of 1981 to the Commission that it could proceed with a civil investigation. . . . At the present time, the Department of Justice is examining the matter to determine whether any NRC official . . . engaged in prosecutable criminal misconduct."[114]

Timothy Martin, an NRC staff member who was involved in the agency's brief 1980 investigation into the Hartman allegations, told the Commission on May 24, 1983, that the staff knew by early 1980 that the leak-rate tests had been falsified.[115] Yet the NRC staff, engaged as a party to the TMI-1 restart hearings on management competence and integrity, did not bring evidence of falsification to the NRC licensing board and instead consistently endorsed the utility's competence and integrity. The staff obliquely informed the board in 1981 that it had identified "problems related to procedure adherence," that no more could be said because the NRC investigation had been suspended at the Justice Department's request, and that information so far obtained indicated "no direct connection with the Unit-2 accident."[116] The board thus regarded the matter as unproven allegations and concluded that "due to our limited information . . . we have no basis to conclude that restart should not be permitted until the DOJ investigation is complete."[117]

The NRC staff later claimed that it had advised the commissioners orally of the seriousness of the leak-rate falsification charges in 1980.[118] Commissioners Victor Gilinsky and Peter Bradford countered that they were not aware of the staff's views on the charges at that time.[119] If the staff privately gave the commissioners information that it withheld from the licensing board, the value of the hearing process is called into question.

In late April 1983, with a grand jury apparently close to indicting Metropolitan Edison, the NRC staff announced that it would have to "revalidate" its position on GPU's integrity.[120] In May 1983, the Commission directed the Office of Investigations to conduct an investigation into the leak-rate charges. In August 1983, the NRC's appeal board reopened the record on the issue of the leak-rate falsifications, but in October 1983 the Commission suspended these hearings until the OI investigation was completed.[121]

Perhaps to avoid interference with the ongoing grand jury probe, the OI did not initially interview any TMI-1 operators. After Metropolitan Edison's indictment and guilty plea in February 1984, the OI was freed from such constraints. However, one month later, the Commission issued a directive that effectively took control of the investigation from the OI and returned it to the Office of Nuclear Reactor Regulation (NRR), the staff division that had endorsed GPU management throughout the restart hearings. The OI was ordered to present all the material it had thus far collected to NRR for "screening." NRR would then refer back to the OI only those matters that NRR decided warranted further investigation.[122]

Not only was overall control of the investigation removed from the OI, but NRR also wrote the official document that constituted the NRC staff's conclusions as to whether the falsification episode as well as a number of other charges raising questions of integrity should disqualify GPU from operating TMI-1. In July 1984, NRR concluded that falsification had occurred. Metropolitan Edison's guilty plea earlier that year left little room for argument on this point. But NRR further concluded that since a "significant change in the licensee's corporate organization" had taken place in late 1981—the change from Metropolitan Edison to General Public Utilities Nuclear—the lack of integrity should not be attributed to GPU Nuclear. The NRC staff thus "reinstated" its endorsement of the GPU Nuclear management.[123] However, both Metropolitan Edison and GPU Nuclear were wholly owned subsidiaries of GPU Corporation; the transfer of the TMI-1 license to GPU Nuclear was performed in a wholly perfunctory manner in August 1981. The board of directors of GPU Nuclear reported to GPU in a typical corporate hierarchical fashion, and William Kuhns was chairman of the board of GPU and all its subsidiaries, as he had been since before the TMI-2 accident.

The United States attorney in Pennsylvania who handled the criminal case was particularly outspoken in his criticism of the NRC's handling of this case. Addressing the U.S. District Court after Metropolitan Edison's guilty plea, he stated:

> We are the only institution since the accident that has made the slightest damn effort to see this thing through to a conclusion. . . . The NRC has not conducted any meaningful investigation; to this day has used as a

pretext the fact that the Grand Jury was conducting an investigation as a vehicle to avoid addressing its responsibilities.[124]

The commissioners rejected the petitions of the state of Pennsylvania and all other intervenors for public hearings to determine whether the systematic leak-rate falsification scheme constituted evidence of corporate dishonesty so pervasive that GPU Nuclear should be disqualified from holding a nuclear plant license. Instead, the commissioners adopted the views of their staff—that prior bad acts should not be imputed to the current management. The Commission authorized restart of TMI-1 without ever permitting the intervenors to question a single witness on this subject, while conceding that leak-rate falsification took place on a daily basis for months and involved the great majority of operators as well as senior operations personnel and management.

The NRC's legal authority to deny hearings was upheld by a 2–1 decision of the U.S. Court of Appeals for the Third Circuit on August 27, 1985. In dissent, Judge Arlin Adams stated:

> It is true that regulatory proceedings can not go on interminably. It is also true that to some extent, the Commission must rely on its licensees in the daily operation of reactors. For just this reason, however, evidence that a licensee has falsified the results of safety test merits special attention. No charge could be more damaging to public confidence in the safety of nuclear power production. The reactor's neighbors must bear the immediate physical and psychological burden of the March 1979 accident. But communities across the nation are looking to the Commission for assurance that they will not be victims of the next accident. They deserve a hearing on serious evidence of a licensee's misconduct before TMI-1 is allowed to restart.[125]

An injunction against operations was issued by Justice Brennan of the U.S. Supreme Court but was vacated by the full Supreme Court on October 2, 1985. The following day, TMI-1 was "taken critical" for the first time in more than six and one-half years. The top managers presiding over its operation were the same persons who presided over the destruction of TMI-2 and the falsification of tests.

Outcome of the OI Investigation of the TMI-2 Cleanup. The NRC established the Office of Investigations largely in response to criticism of problematic NRC staff investigations.[126] Such criticisms had come from Representatives Morris Udall, Toby Moffett, and Edward Markey, among others. Since the NRC staff acts as a full party to licensing cases and generally supports the utilities' requests for licenses, there is an inherent conflict in expecting it to vigorously pursue criminal investigations of the utilities. The OI reports directly to the Commission and is not part of the NRC staff, which reports to the executive director for operations. This distinction is central to the OI's ability to function independently.

The OI appears to be a markedly better investigator than was the Office of Inspection and Enforcement. Nevertheless, there is a serious question as to whether the OI has the personnel needed to do its job. It has reportedly told the Commission on a few occasions of its need for more staff, but it has been unsuccessful in obtaining this needed increase.[127] According to the NRC's Office of Congressional Affairs, the Commission submitted a reprogramming letter in 1983 asking the House Appropriations subcommittee that controls the NRC's budget for additional OI staff, but the request was not approved.[128] (The subcommittee reportedly objected that the NRC had not responded to its earlier request for an explanation of the difference between the Office of Investigations and the Office of Inspection and Enforcement.[129]) The NRC came back with a new request—cut in half this time—but the subcommittee again would not approve it. The Commission has not asked for additional OI staff in its subsequent budget requests.[130]

Perhaps more importantly, events since the OI's establishment cast doubt on the degree of the new office's freedom from pressure from other sections of the agency whose purposes often conflict with the OI's. As in the leak-rate falsification case, the OI's effectiveness was called in question during the TMI-2 cleanup operation. In March 1983, engineers working on the cleanup alleged that Bechtel Corporation and GPU had engaged in quality-assurance violations in preparing a crane that was to lift the head off the reactor and allow access to the damaged fuel. In September 1983, the OI issued a report that not only substantiated these allegations but also found that they were "illustrative, not exhaustive." The OI also found that the problems had been aggravated by the NRC staff's approval of Bechtel procedures and work that did not meet GPU's requirements.[131]

The Commission directed the NRC staff to review the OI report and suggest a plan for corrective action. The staff agreed with certain of the findings but said the problems were not significant for safety and there was "no evidence of deliberate circumvention of procedures." The staff told GPU that no fine would be issued, in part because of the lack of evidence of deliberateness. The staff also disagreed with the OI's finding that it had contributed to the problems.[132]

Then, one month after the crane had lifted the reactor's head, cleanup workers discovered that a mechanism had malfunctioned, disabling one of the crane's two main brake systems and threatening the other. GPU had added the mechanism to the brakes in 1982 but never reported it to the NRC, and it had escaped quality-assurance review. After this event, the staff met with the OI to discuss why the OI continued to believe there was evidence of deliberate violations in 1983. On October 29, 1984, the staff acknowledged that "the circumvention of requirements was at least to some degree deliberate, and their motivation appeared to be expediency not confusion." The staff requested that the OI's views "supersede the relevant staff views expressed previously."[133]

The House Appropriations subcommittee that controls the NRC's budget, chaired by Representative Tom Bevill, has stated that it was "not convinced that there is any need for a separate Office of Investigations."[134] Unfortunately, the Commission reportedly was considering making the OI part of the NRC staff, a move that would mean a leap backward from the progress the Commission made in establishing the OI.

Staff Investigation of Material False Statements at Diablo Canyon. This case involved the adequacy of an NRC investigation of false statements by Pacific Gas and Electric Company concerning an "independent" report prepared by R. L. Cloud Associates, drafts of which PG&E reviewed and commented on. After PG&E discovered errors in the seismic design diagrams for Diablo Canyon Unit 1, it asked Cloud to determine whether other errors had been made in the plant's seismic design. The NRC requested that it be furnished with a report on the results of Cloud's review prior to fuel loading.

On October 21, 1981, the Cloud firm submitted a draft of the report to PG&E. Four high-level PG&E officials received copies. PG&E then gave comments to Cloud, which prepared a revised draft of the report. On October 26, the revised draft was given to PG&E.[135] PG&E and Cloud officials also met to prepare for a meeting they were to have with the NRC staff on November 3. At the November 3 meeting, the NRC staff asked whether and when the NRC would get the Cloud report (the same one PG&E had already received) and discussed the necessity of preserving the report's independence. Present at the meeting were R. L. Cloud, PG&E Vice-President George Maneatis, and a PG&E attorney, Bruce Norton. No one at the meeting disclosed that PG&E had seen and prepared comments on the drafts; in fact, Cloud, Maneatis, and Norton made statements that led the NRC to believe that PG&E did not have the drafts.[136] While at least six persons at the meeting knew of the drafts, no one corrected the record during or after the meeting.[137] Three days after this meeting, the Cloud firm submitted a third draft of the report to PG&E. On November 12, the final report was sent to PG&E, and on November 18, PG&E sent a copy to the NRC. The NRC finally received the Cloud report on November 25.[138]

The NRC staff learned of the existence of the October 21 draft from the House Committee on Interior and Insular Affairs in early December and eventually obtained PG&E's copies of this draft along with the utility's comments. In mid-December, Representatives Morris Udall and Leon Panetta expressed concern to the Commission about the Cloud affair,[139] and Dircks, the NRC's operations director, ordered Region V to conduct an investigation.

The investigation, finished in approximately one month, was superficial and limited. According to a paper prepared by the Interior and Insular Affairs Committee staff and sent to the Commission by Udall:

> pertinent questions were not asked. Not only was no attempt made to resolve important discrepancies in statements to the investigators, but the fact of such discrepancies was not highlighted either in [the Staff's investigation report] or in presentations before the Commission.
>
> In sum, the Region V inquiry leaves the impression that the NRC investigators sought to avoid collecting information that might have led to a finding of willfulness.[140]

The House committee's staff found about eighteen significant deficiencies in the NRC's investigation.[141] For example, the investigation report did not indicate the nature of the discussion that took place at the Cloud/PG&E meetings held just prior to the November 3 meeting with the NRC staff. Since these meetings concerned what would be presented to the staff (that is, the Cloud findings), it seemed likely that the Cloud report would have been discussed. Yet some PG&E representatives at these meetings later maintained that they were not aware that the draft had been given to PG&E. The content of the Cloud/PG&E meetings could therefore be expected to shed some light on the credibility of these denials. Referring to statements made at the November 3 meeting by Maneatis, Norton, and Cloud, a PG&E supervising licensing engineer said: "I had some I guess you'd call it anxiety that I was not sure if those statements were totally in line with the fact that the presentation was made by Dr. Cloud to PG&E that weekend."[142] The NRC's investigation report did not address these potential discrepancies.[143]

A second deficiency was the investigation's failure to discuss the fact that PG&E did not give the NRC all the draft reports when the NRC

asked for them. PG&E first turned over only one copy of the October
21 draft; it did not inform the NRC of the other copies that contained
PG&E comments, nor of the October 26 and November 6 drafts. When
the NRC found out about the additional copies of the first draft, PG&E
handed them over but still failed to mention the two later drafts.[144] A
third problem was the investigation's failure to probe whether one Cloud
associate's belief that PG&E's comments made the Cloud report less critical of the company was shared by others at Cloud. That would have
helped determine whether Cloud or PG&E had a motive to conceal that
PG&E had commented on the draft.[145]

The deficiencies in the NRC's investigation report were significant.
The unasked questions might have shed light on the degree to which
misleading or false statements made by Cloud and PG&E at the November 3 meeting (and their subsequent failure to correct these statements)
were intentional. Answers to these questions should have influenced the
disciplinary action taken by the Commission. On February 10, 1982, the
Commission voted (John Ahearne and Thomas Roberts dissenting) to
issue PG&E a notice of violation for making a material false statement
prohibited by the Atomic Energy Act, but no fine was levied. (A fine is
the NRC's standard disciplinary action when a licensee commits a significant violation of the rules.) The commission merely directed the staff
to meet with PG&E to discuss the problem and stated that the Commission intended PG&E to take unspecified steps to remedy the problem.[146] As the Interior and Insular Affairs Committee staff observed,
the deficiencies in the NRC investigation "have befogged the significance" of the Commission's finding that PG&E had made a material false
statement:

> on the one hand, a reading of the summary of [the NRC's investigation
> report] and the Commission's February 10 statement could lead to an
> inference that the material false statement was a result of an innocent
> misunderstanding. On the other hand, analysis of the entire record sug-
> gests that [PG&E's actions] might have constituted a willful effort to con-
> ceal the nature of the Cloud/PG&E relationship as it existed prior to
> November 3.[147]

Commissioner Gilinsky pointed out that the Commission's failure
to levy a fine undermined the seriousness of the offense.[148] Representative Richard Ottinger similarly criticized the Commission for issuing a
notice of violation "unaccompanied by any meaningful sanction."[149]

The Commission's February 10 order also stated that it was keeping
under review PG&E's nomination of R. L. Cloud Associates to be the

primary auditor for its major design reverification program. Yet Cloud had demonstrated a lack of independence, not only in giving PG&E the draft reports on which to comment but also in later failing to tell the NRC about it. Commissioner Gilinsky and Representatives John Dingell and Ottinger castigated the Commission's failure to reject Cloud as a candidate for reviewer of Diablo Canyon's safety.[150]

In February 1982, Representative Udall called for an OIA review of the staff's investigation.[151] Two and one-half months later the Commission ordered one. The OIA ultimately found "basic and substantive flaws" in the staff's investigation. For example, it criticized the fact that "the overall scope of the investigation did not include obtaining the testimony of key NRC officials who were intimately involved and had knowledge regarding key issues under investigation."[152] The OIA also criticized the weakness of the staff's interviews. For example, an NRC investigator asked PG&E attorney Norton whether he had known, when he made certain statements at the November 3 meeting, that draft reports had been given to PG&E for review and comment. Norton answered, "I did not know that," and, according to the OIA, "the interview was then for all intents and purposes ended. The entire interview of Norton was conducted in approximately 10 minutes and clearly made no attempt to probe the question under consideration."[153] In response to questions from Udall, the OIA concluded that "the questioning of witnesses by NRC staff was in many instances neither sufficiently aggressive nor probing. Further, several opportunities to pursue viable lines of questioning, based on the responses received, were not followed up. Accordingly, minimal investigative results were derived from the overall interview process."[154]

The original report of the OIA's senior criminal investigator, Ronald Smith, has not been made public. However, according to a congressional subcommittee staff memorandum, Smith concluded that the NRC investigation "did not represent a professional investigative product"; it was "not adequate and at best represented only an amateur effort."[155]

In spite of the OIA's findings, the Commission decided not to reopen the investigation, and no disciplinary action was taken against the NRC staff for its shoddy work. The Commission eventually approved the use of Teledyne Engineering Services to handle the design verification program at Diablo Canyon. Teledyne then hired R. L. Cloud Associates, which ended up playing a major role in the safety review.[156]

Investigation into Withholding of Information. During the tense days of the accident at Three Mile Island, significant information pertaining to the accident's severity was not communicated to Pennsylvania officials by the management of Metropolitan Edison, the TMI licensee. Because the utility's authority exists only on the plant site, any off-site actions to protect the public generally must be undertaken by state or local officials. An investigation into this matter, carried out by the House Committee on Interior and Insular Affairs, reached the following conclusions:

> The record indicates that in reporting to State and Federal officials on March 28, 1979, TMI managers did not communicate information in their possession that they understood to be related to the severity of the situation. The lack of such information prevented State and Federal officials from accurately assessing the condition of the plant. In addition, the record indicates that TMI managers presented State and Federal officials misleading statements (i.e. statements that were inaccurate and incomplete) that conveyed the impression the accident was substantially less severe and the situation more under control than what the managers themselves believed and what was in fact the case.[157]

Metropolitan Edison's failure to communicate information during the accident appeared not to trouble the NRC, which did not include the issue in NUREG-0600, its 1979 accident investigation report. Finally, after prodding from the House committee (with each step "taken reluctantly and grudgingly," according to Commissioner Gilinsky[158]), the NRC's Office of Inspection and Enforcement undertook an investigation. The IE report, completed in January 1981, contained conflicting conclusions. It found both that "Met Ed was not fully forthcoming on March 28, 1979 in that they did not appraise [sic] the Commonwealth of Pennsylvania of either the uncertainty concerning the adequacy of core cooling or the potential for degradation of plant conditions," and that "information was not intentionally withheld" from the state and the NRC. IE recommended that no citation for failure to report be issued.[159]

David Gamble, a former criminal investigator with the Office of Inspector and Auditor who was assigned to the IE investigation to protect the interests of the Justice Department in criminal matters that might arise, revealed later that both the investigators and the scope of the investigation were severely restricted. In a January 1981 memorandum to Norman Moseley, the investigation coordinator, and in testimony submitted under oath for NRC hearings on GPU's competence and integrity in November 1984,* Gamble described a number of investigative practices he considered inappropriate. For example:

---

*At the time Gamble prepared this testimony, he had become a supervisory criminal investigator with the Defense Criminal Investigative Service.

Two members of the investigative team, Ronald C. Haynes and William L. Fisher, both of IE, drafted portions of the report for which they were responsible prior to conducting any interviews. It is my opinion that writing sections of the report before engaging in any significant investigation of the facts indicated that they may have predetermined the conclusions they would reach.[160]

Another practice of the principal NRC personnel assigned to the case, according to Gamble, was to draft "the questions to be asked during all interviews" and then try to "prohibit other interviewers from asking questions outside the pre-approved list, even those flowing logically from the witnesses' answers." Other interviewers wishing to ask additional questions had to wait until an interview was finished and then ask Moseley's permission. Moseley then screened the questions and determined whether they could be asked "on the record."[161]

Gamble also stated that NRC investigators agreed in advance with attorneys for at least one company on the topics about which employees would be questioned and that the company's lawyers were permitted to be present during interviews, which "may have had a chilling effect upon interviewees and provided opportunities for improper coaching of later interviewees to provide consistent testimony."[162] Gamble believed that the NRC investigation should have been headed by "trained investigators using standard investigative techniques instead of by IE inspectors and IE management."[163]

According to Gamble, Moseley prohibited the investigators from asking about Metropolitan Edison's communications with state officials because Moseley "did not want to ask questions which might provide Pennslyvania authorities a forum in which to complain that they had not received sufficient information from the licensee."[164] In a January 1981 memorandum, Gamble told Moseley that the NRC investigation report drew conclusions about information flow to the state when this issue was not satisfactorily investigated. For example, IE did not interview Pennsylvania Lieutenant Governor William Scranton because of schedule conflicts but made no attempt to discover whether any minutes were taken of the utility's briefing of him during the accident. Gamble concluded: "It seems strange now to take action based upon this restricted phase of the investigation."[165] Gamble also wrote Moseley:

Throughout the report are conclusions which I do not feel are adequately supported by the report. While the opposite conclusion would not be justified either, the report confuses opinions with conclusion—implicit in the latter is that they have a factual basis. For example . . . you conclude that none of the conflicts examined were the results of lying; however, it is just as reasonable based upon the facts presented in your report to conclude that they were the result of lying.[166]

Gamble later stated that he recalled IE officials discussing

which word would be most appropriate to describe the intent of the licensee's actions. In general, IE appeared to want to avoid use of the word "willful" even though words connoting similar intentional or deliberate action were considered. I believe part of the reason IE wished to avoid use of the word "willful" was because of a desire not to indicate the company's potential exposure to criminal liability for its actions.[167]

The NRC staff's efforts to avoid the legal implication of the term "willfulness" is shown in IE Director Stello's characterization of the utility's action as withholding information from the state and the NRC "knowingly" but not "willfully."[168] Other NRC staff descriptions included "not fully forthcoming" and "dissembling"; one IE official is purported to have stated that the company did "not level" with the state officials.[169] The commissioners attempted to understand Stello's distinction and questioned him closely on the meaning of the words, pointing out that they could imply intent. After hearing the dictionary definition of *dissemble*, Stello said: "Well, then, I will invent a new word."[170]

With the exception of Commissioners Gilinsky and Bradford, the Commission defended IE's position. The NRC issued Metropolitan Edison a notice of violation with no response required. Answering a letter from Representative Udall, Gilinsky charged that

just as the Company was too weak to level with the State and Federal authorities on the day of the accident, and too weak to confront that failure afterward, so NRC has been too weak to carry out its responsibilities. Both the Commission and the Staff have hidden behind every ambiguity to explain away any wrongdoing connected with the reporting failure.[171]

Udall also expressed his discouragement with the Commission's failure to demand integrity from its licensees:

Recently we have seen the sorry spectacle of high level Commission staff seeking to make a distinction between "knowingly" and "willfully" withholding information . . . a distinction that appears acceptable to a Commission majority.[172]

Quality-Assurance Breakdown at Zimmer. The painful saga of construction problems at the Zimmer plant involved a litany of NRC bungling and misconduct that helped drive the plant to an early death and turned a spotlight on the problem of lax regulation by the agency.* This litany included the NRC staff's failure for four years to systematically investigate construction deficiencies despite warnings from the NRC's resident inspector, a half-hearted NRC staff investigation, another NRC staff investigation that downplayed the seriousness of Zimmer's problems, the belated acknowledgment by high-level NRC officials that the problems were serious (after large-scale public and congressional pressure and the evidence of problems proliferated), an OIA investigation report that was restricted in scope and omitted evidence of the NRC's and Cincinnati Gas and Electric's early awareness of problems, and the OIA's obstruction of the release of documents in response to a citizen's request under the Freedom of Information Act. While CG&E, the plant's lead owner, must take most of the blame for Zimmer's breakdown and eventual death, the plant's history indicates that if the NRC had not for years closed its ears to massive evidence of serious safety problems, Zimmer might be operating today.

Investigation of the breakdown uncovered serious questions about the quality of thousands of welds, plant wiring, inadequate or non-existent weld inspections, unqualified welders, steel beams of indeterminate quality, scores of construction records that were falsified or destroyed, and verbal and physical harassment and intimidation of plant inspectors who attempted to call attention to these problems.[173] Evidence of problems was first brought to the NRC's attention in 1976 when Victor Griffin, a quality-assurance engineer working for the utility's major contractor, Henry J. Kaiser Company, quit his job because of what he said were serious safety violations at the plant. The NRC sent in investigators, but within a few days they concluded that there was no serious problem.[174]

The next person to blow the whistle was Terry Harpster, a reactor startup specialist for the NRC and principal inspector at Zimmer from 1977 to 1979. Harpster later told NRC investigators that his inspections documented a number of problems and that he had concluded the plant was "out of control." Harpster said that he was able to set up a meeting between the NRC and the utility to discuss problems at the plant only

---

*Zimmer was canceled as a nuclear plant in January 1984 after its owners determined they could not afford the necessary rework. A total of $1.7 billion had been spent and the plant was described as 97 percent complete, but it was estimated that to satisfactorily reinspect and finish the plant (even without making significant repairs) would nearly double the price tag.

after he "screamed" at NRC licensing officials. Little came of this meeting. Harpster criticized the NRC's licensing process, stating that Nuclear Reactor Regulation officials responsible for licensing were on "a very tight schedule" and that "the IE inspector is often viewed by NRR as an adversary when he uncovers deficiencies which NRR has already 'blessed.' "[175] The NRC did not make Harpster's statements public until August 1982.

In 1979, a private detective, Thomas Applegate, arrived at Zimmer on a divorce case and discovered time-card fraud. When he gave information on the fraud to CG&E, the company hired him to investigate misconduct of Zimmer workers. In talking to the workers, Applegate heard allegations of problems with the plant's welds and piping. CG&E was not interested in these allegations and terminated Applegate's employment. Applegate went to the NRC.[176]

In early 1980, Applegate repeatedly attempted to contact NRC officials, primarily the director of the OIA, James Cummings. Most of his attempts were unsuccessful. When he did speak with Cummings, Applegate reported receiving a lukewarm response. Finally, he went directly to John Ahearne, who at that time was the NRC chairman. Had Applegate not done so, "potentially significant information could have been unnecessarily delayed or lost to the Commission's regulatory program," according to the Hoyt report on an internal NRC investigation of the OIA's handling of Applegate's allegations.[177] The NRC eventually sent investigators, who in July 1980 concluded that there were no safety problems at Zimmer. [178]

Applegate, concerned over NRC practices in investigating his allegations, contacted the Government Accountability Project (GAP), an organization that represents industry and government whistleblowers. In December 1980, GAP took Applegate's allegations to the U.S. Merit Systems Protection Board. As a result, the Commission directed its IE office to reopen the Zimmer investigation.[179] The Commission also ordered an internal inquiry by the OIA.

The OIA review, completed and sent to the Commission in August 1981, criticized the IE investigation as "unsatisfactory" because it "failed to properly document" its results and "was neither vigorous nor sufficiently broad in scope." Specifically, it "failed to determine the correct status and history of several welds," and its "finding of 'not substantiated' with regard to the allegation that defective welds in safety-related systems had been accepted is not consistent with the facts."[180] For example, of three welds that Applegate had alleged were defective, two were repaired or replaced shortly before or during the IE investigation. IE neither identified the late dates on which this work was done nor mentioned that one weld, which had previously been marked as accepted, was cut out and repaired just before the IE investigation. Interviews revealed that IE investigators had no idea that the other weld had been replaced. In fact, the investigators had not inspected any of the welds in question or reviewed all the pertinent documentation.[181] Nevertheless, the OIA did not make any "corrective recommendations," and the agency did not take disciplinary action against any NRC staff members responsible for the faulty IE investigation.

During the reopened IE investigation into problems at Zimmer, IE investigator James McCarten received scores of new allegations (about 90 in a two-week period) from plant workers. Many of the allegations had possible criminal implications. McCarten reported later that he attempted to get his superiors to pursue the allegations, believing that the NRC could potentially build a criminal case, but "they did not want to have anything to do with a criminal investigation."[182] During this time, Regional Administrator James Keppler circulated within the NRC an article by the Ethics and Public Policy Center that attempted to discredit GAP. The article described GAP as part of a pro-Soviet movement, supporting "Third World terrorists" and using "guerrilla politics" aimed at "the destruction of the capitalist system."[183] When NRC inspectors began to review the new allegations, they confirmed problems so consistently that they recommended that construction be stopped immediately. Region III Administrator James Keppler rejected this proposal. In a 1983 interview, McCarten recalled that Keppler had responded: "[H]ow can you guys tell me that a plant is a hundred percent inspected, every construction module was bought off on . . . and yet you are telling me it is not built right. How did that happen? . . . . how can I go to the public and say we have complete[ly] inspected the plant, it is 93 percent complete, but it is a mess?"[184] Keppler later denied this motive for refusing to halt construction, insisting that what was seen at the plant at that time was "a management control problem," not serious "hardware" problems.[185]

It was Keppler's Midwest Region III that was responsible for overseeing Zimmer, Marble Hill, and Midland, all of which were canceled, largely because of delays due to quality-assurance and hardware problems. At Midland, as at Zimmer, Region III approved plans initially favorable to the utilities but ultimately contributing to the plant's insurmountable problems. Region III approved Consumers Power's plan to resolve a problem of sinking foundations at Midland by filling the sinking and cracked building with sand, in spite of warnings from some NRC engineers that the plan was inadequate.[186] In June 1983, a Region III soils expert, Ross Landsman, said the foundation work the NRC had approved would not completely solve the problem.[187] In October 1984, NRC inspectors recommended issuing the utility a $100,000 fine for excavating in an area expressly prohibited by the NRC.[188] However, after a meeting in a coffee shop at Chicago's O'Hare Airport between a Consumers Power attorney and the NRC's director of inspection and enforcement, Richard DeYoung, the fine was not issued. Instead, the utility was ordered to conduct a management audit. Keppler said he was not pushed into dropping the fine.[189]

Ironically, Keppler is considered by many to be one of the tougher NRC regional administrators. Region III inspector Isa Yin was anxious to return to his region after experiencing Region V practices while on detail to the Diablo Canyon plant. (See the section "Lessons Lost from Zimmer: The NRC Response to a Quality-Assurance Breakdown at Diablo Canyon" in this chapter.) Commissioner Gilinsky rated Keppler, along with Southeast Region II Administrator James Reilly, as the most effective of the five regional administrators. Gilinsky said that Keppler "has, on occasion, come in with strong recommendations and been talked into softening them by people above him." One former NRC staff member reportedly said that "Jim Keppler is the brightest spot in the country—and that'll tell you how bad the NRC is."[190]

At a meeting in May 1981—after receiving briefings by NRC personnel on improper welds and electrical installations, lack of documentation for materials, harassment and intimidation of inspectors, and voiding of hundreds of reports on construction problems—IE Director Stello also maintained that the problems were "paperwork" and not hardware.[191] According to McCarten, when he described criminal falsification of records that hid hardware problems, "Stello got very emotionally upset, threw up his hands and said we have got inspectors tied up in grand juries right now. I don't want to hear about any criminal allegations. He says we are just going [to] do health and safety. That is OIA's job and we don't want to have nothing to do with criminal stuff, and this is May. Then he walked out of the meeting."[192] (Three participants in this meeting recalled Stello saying that just by "looking around" a nuclear plant, an NRC inspector should be able to know immediately whether it was "well built or not." According to one investigator, there was "much guffawing after the meeting about the possibility of firing the inspection staff and just putting Mr. Stello on a plane and just flying him around to nuclear power plants to look around."[193])

Similarly, the Hoyt report found that Region III Division Director Charles Norelius' statement that "I did not view our primary purpose to be one of focusing on wrongdoing per se" was "reasonably consistent with those articulated by senior IE officials."[194] The report concluded that "Victor Stello . . . has taken the official position and did so instruct Region III officials in 1981 that IE personnel are not to conduct criminal investigations. While facially correct, we further find that Region III officials apparently relied upon Mr. Stello's leadership on this aspect of enforcement to justify a failure to vigorously pursue *all* possible causes or types of regulatory violations."[195] Stello later repeated that it was Commission policy that IE only "focus on the health and safety implications of regulatory noncompliance," that criminal issues were left to the OIA and the Department of Justice, and that the commissioners were informed that IE was not conducting a criminal investigation.[196] However, as noted by the Justice Department, the OIA's "knowledge of potential cases was dependent upon referrals" from IE.[197]

The Hoyt report further concluded that both the NRC and the nuclear industry would suffer from such a view of the NRC's enforcement obligations:

> IE's hesitancy to pursue all potential causes, criminal or otherwise, for the altered QA/QC [quality assurance/quality control] documents it had identified contributed to the public perception of impropriety and inadequacy with respect to the NRC's response to Zimmer. In our view, an identification of QA/QC deficiencies is but the first step in ensuring compliance with the Commission's regulatory requirements. The necessary second step is a full exploration of the potential causes of such deficiencies. To do the former but not the latter is, in our view, akin to treating only the symptoms of a cancer. In our mind, the results in each case will be the same—a longer period of illness ending in the death of the patient.[198]

In November 1981, the NRC issued an interim investigation report that found widespread quality-assurance violations at Zimmer and fined the utility $200,000. While the report summary acknowledged "a few" hardware problems, it maintained that the problems found were "largely programmatic."[199] While emphasizing the need for the ongoing "quality confirmation" program, the NRC allowed construction to continue.

According to IE investigator McCarten, however, significant findings of the investigation were deleted, edited, or underplayed by Region III officials in producing the November 1981 report. McCarten said that

IE officials told him after seeing portions of the report that they would have to "use word engineering" and "massage" it. According to Mc-Carten, the report was edited to the point where it was "so watered down and . . . so toned down that . . . the impact is gone." McCarten also said that parts of the report given to the commissioners were "false and contradicted sworn depositions that were attached to the same report."[200] McCarten said that he protested orally and in a memorandum the editing done by Region III officials.[201] Keppler later responded to McCarten's charges by claiming that McCarten had been given the opportunity to review the report at various stages and had agreed to the changes made.[202] (In 1982, McCarten resigned from the agency to take a job as a special agent for the Naval Investigative Service. He said he left the NRC because "regional officials of the NRC were not adequately doing their job in the construction inspection program and were quashing any information which would prove that."[203])

The Commission's decision to suspend construction came only after new evidence of serious problems released by GAP, widespread opposition to continued construction, a media barrage that included an investigative series by Gannett News Service, hearings by the Cincinnati Environmental Advisory Council, hearings by a House Interior and Insular Affairs subcommittee, the disclosure of a grand jury investigation, and Keppler's announcement to the Commission that significant hardware problems were being discovered and that allegations were coming in faster than they could be dealt with. In November 1982, with Roberts and Ahearne dissenting, the Commission voted to suspend construction.

An investigation by federal prosecutors into activities at the Zimmer site was also delayed by the NRC. In June 1981, OIA field investigators determined that they had "a strong indication" that CG&E or Kaiser Company management had subverted the quality-assurance program at Zimmer; they felt the matter should be referred to the Department of Justice. OIA Director Cummings opted instead for an OIA-conducted criminal investigation (where a case is developed and then referred to the Justice Department). The OIA committed itself to provide the Justice Department with an analysis of the potential criminal violations, but the analysis never came. This experience "clearly affected the credibility of the NRC in dealing with the Department of Justice," according to the Hoyt report.[204] Anne Tracey, the assistant U.S. attorney in charge of the Zimmer case, said that getting reports and assessments from the OIA was "like pulling teeth." Tracey said that the Justice Department was "delayed probably a good year" in beginning its criminal investigation:

"OIA was supposed to be the liaison with us and we weren't getting anywhere ... so we were always kind of waiting and twiddling our thumbs till we finally figured out that we could not rely on OIA before we took some action,"[205] The Justice Department began to investigate in June 1982, and at this writing its probe was still going on.

It was revealed through a Freedom of Information Act request filed by Applegate and GAP that the August 1981 report by the OIA, while critical of IE's earlier investigation, had been edited by Cummings to omit significant information. The OIA investigators had defined their task as reviewing the IE investigation and then determining whether the deficiencies discovered were the result of inadequacies or weaknesses in Region III's investigation procedures. In editing the OIA report, however, Cummings redefined its scope. Over the objections of two of his principal investigators, he restricted it to a review of the IE investigation of the Applegate allegations, not examining whether this review had disclosed broader problems with Region III generally. Helen Hoyt, the Atomic Safety and Licensing Board administrative law judge who conducted the internal NRC investigation into Applegate's allegations, concluded that the original scope would have been more valuable to the NRC and that the agency was "little served by the artificially limited conclusions of OIA's August report."[206]

In narrowing the scope of the report, Cummings deleted a section[207] containing evidence that both the NRC staff and top CG&E executives were aware of serious problems at Zimmer as early as 1977–79.[208] The OIA report also failed to document that the interview revealing this evidence even took place. According to Hoyt, that was contrary to standard investigatory practice. The Hoyt report found "the failure to do this particularly troublesome because the interview contained information that appeared to be damaging to both the NRC and CG&E. An office that is intended to be the equivalent of an Inspector General's Office should maintain practices that are beyond even the appearance of impropriety. We must conclude that Cummings' handling of the Harpster interview did not satisfy that standard."[209]

Before the OIA report was completed, Cummings ordered two of his investigators to review the final draft with Region III IE officials, whose conduct the report was evaluating. One investigator strongly objected to Cummings's order and the other expressed reservations. Cummings then made changes in the report to, in his words, "meet Mr. Keppler's requirements." Hoyt found Cummings's action to be "*highly improper* investigatory conduct":

Every investigator and inspector (except Mr. Cummings) questioned indicated that the showing of a draft report to its subjects prior to its release
was unusual and improper investigatory practice. Yet Mr. Cummings
stated that although it was a close decision on his part and not a "normal
thing to do," he does not think it was "a wrong thing to do," nor does he
regret it.[210]

Hoyt also determined that Cummings had attempted to conceal his
changes in the report by directing that no written record of them be
kept. After Cummings signed the report, however, OIA investigator
John Sinclair disobeyed Cummings's orders and identified the changes,
saying that Cummings's directions would be something he could not "live
with" and could be an "absolute catastrophe."[211]

Finally, Cummings attempted to thwart the release of Zimmer documents requested under the Freedom of Information Act (FOIA). When
Applegate and GAP discovered that the August 1981 OIA report had
been significantly edited, they requested, pursuant to the FOIA, all background documents pertaining to the OIA review. The NRC released
some documents, but it neither released nor denied (that is, it did not
acknowledge the existence of) any drafts or informal records. In responding to an appeal, the NRC maintained that no such documents
"were retained in NRC files." But after Representative Udall began an
investigation into NRC record-keeping procedures and after GAP and
Applegate sued the NRC, the agency released numerous documents that
it had previously denied existed.[212]

In December 1982, OIA investigator Gamble informed Udall that
Cummings had directed Gamble to remove documents from OIA offices,
with the result that they were not listed in response to the Applegate
FOIA request.[213] The Commission later told Udall that Cummings had
a policy directing that drafts and background documents not "essential
to understanding of final [investigation] reports" be removed from the
investigation files and the agency premises. The Commission said that
Cummings felt that keeping such documents "was wasteful from the
standpoint of clerical filing time and reduced storage capability [and]
served no useful purpose. . . ." The Commission explained that some
documents had not been identified because, under Cummings's policy,
they had been ordered removed from the agency premises and thus
"were no longer 'agency records' subject to the FOIA. . . ."[214] (Chairman
Palladino later sent a memorandum to all NRC employees, instructing
them that drafts, background materials, and notes could be destroyed
unless an FOIA request had already been made for them—in effect,

authorizing a preemptive sanitization of the agency's files.[215]) Other documents were not mentioned because they were part of another OIA investigation of construction defects at Zimmer "and thus were not within the scope of [Applegate's FOIA] request." One such document was the interview with Harpster, although the Commission acknowledged that there was "no dispute that the Harpster interview should have been identified."[216]

The U.S. District Court was less understanding. In its May 1983 decision on the Applegate FOIA lawsuit, it found that

> evidence was uncovered in the record suggesting that despite the existence of carefully drafted official NRC FOIA policies and procedures, the personnel assigned to implement FOIA in OIA executed those rules in a manner designed to thwart the release of responsive materials. These procedures appeared to include the removal of documents from agency files, taking documents home, and the use of carefully worded oral inquiries designed to avoid identification of documents.
>
> It is disturbing to this Court that unbeknownst to agency management, an office in the NRC was able to design a filing and oral search system which could frustrate the clear and express purposes of FOIA. The assertion of an exemption is one thing, avoidance borders on dishonesty. It is also disturbing that FOIA appears to have been implemented in an adversarial manner. A lawsuit ought not be required to ensure the adequacy of a search.[217]

The actions the NRC took toward individuals implicated in the Zimmer case were almost as disturbing as the poor practices themselves. In response to the Hoyt report's criticisms of Stello, Chairman Palladino wrote to Stello to "carry out the wish of the Commission that you receive a clear expression of our continued support and confidence."[218] (Julian Greenspun of the Department of Justice also noted, presumably referring to Stello, that prior to mid-1982 "the person in charge of [IE] was not anxious to have criminal cases developed and, indeed was somewhat protective of the operators of nuclear power plants."[219]) In 1982, Stello was promoted from IE director to deputy executive director for regional operations and generic requirements, the second-highest NRC staff position. From 1981 to 1983 he received annual senior executive service bonus awards totaling more than $29,000.[220] Cummings was removed from his post as OIA director in 1983 but was soon granted an eleven-month sabbatical, during which he attended the University of Virginia while receiving his full NRC salary, tuition, and a tax-free per diem that may have totaled as much as $9,300.[221] He subsequently returned to the

NRC. With rewards such as these, the Commission has sent few signals that would discourage the recurrence of the practices that occurred at Zimmer.

Lessons Lost from Zimmer: The NRC Response to a Quality-Assurance Breakdown at Diablo Canyon. Unfortunately, the NRC appears to have learned the wrong lesson from Zimmer. Rather than taking seriously the evidence of problems at Diablo Canyon alleged by plant workers and its own staff, the NRC took an even more adverse position toward whistleblowers' allegations in a rush to license the plant. Diablo Canyon was removed from the list of plants facing licensing "delay" and was allowed to operate three miles from a major earthquake fault while the adequacy of its seismic design remained in controversy.

The NRC did not face up to mounting evidence at Diablo Canyon of a lack of adequate quality assurance, which intervenors had tried for years to bring to the Commission's attention. Finally, after a low-power license was granted in September 1981, PG&E's discovery that reversed seismic diagrams had been used in installing supports for Unit 1's equipment provided undeniable proof of a serious breakdown in the utility's quality-assurance program. After disasters at Zimmer, South Texas, and Diablo Canyon indicated that the NRC's quality-assurance program might bear improvement, the OIA attempted to perform a review of the program, informing Executive Director for Operations William Dircks in October 1981 that its primary focus would be the quality-assurance activities of the NRC staff.[222] The next month, the Commission put the OIA on hold until it had considered the audit.[223] The OIA pleaded its case to the Commission, describing its intention to "ascertain whether or not [the Office of Nuclear Reactor Regulation] and [the Office of Inspection and Enforcement] have adequately performed the tasks they are committed to do by their own established programs."[224] Chairman Palladino decided instead to ask Dircks to review the situation. Dircks was responsible for the NRC offices and staff that were to be the subject of the OIA's investigation. According to a memo prepared by the staff of a House Interior and Insular Affairs subcommittee, "It would appear reasonable to conclude from this chain of events that the Commissioners quashed an 'independent' review of the NRC's quality assurance programs in favor of internal evaluation by the very people who were responsible for formulating and enforcing the agency's regulations."[225]

The resultant reanalysis of seismic structures and components, including an independent design verification program established to verify that the seismic design problems had been identified and corrected, was itself riddled with new problems. As these problems surfaced, additional questions about construction quality assurance were also identified. In 1984, more than 1,000 allegations of quality-assurance problems were submitted by the intervenor (San Luis Obispo Mothers for Peace), with supporting documentary exhibits and sworn statements from plant workers.[226]

In January 1984, only days after the announcement of cancellations at Zimmer and Marble Hill caused largely by quality-assurance breakdowns and at a time when the NRC staff was receiving numerous allegations of problems from whistleblowers at Diablo Canyon, Chairman Palladino expressed to the staff his apparent disturbance over the "flood of allegations" at the "last minute," just as the licensing decision was about to be made.[227] PG&E wanted permission to begin low-power operation that spring. The NRC called the allegations "late" even though, according to the whistleblowers' attorneys, the plant workers had in some cases spent more than two years trying to obtain corrective action from PG&E. In March 1984, Palladino called for a permanent plan for "handling last-minute allegations." He suggested that "such a policy might include a deadline after which the threshold for allowing allegations to hold up a licensing action is very high."[228] In August 1984, the staff produced a plan that found that the method of handling late allegations at Diablo Canyon had "provided a workable framework" and suggested that it be used as a model for other cases.[229] Instead of achieving the corrective action, plant workers said, they were intimidated, harassed, or fired.[230] The staff argued that most of the allegations were not significant for safety or were not new and referred many of them to PG&E rather than to the NRC staff for review.[231]

In March 1984, as the Commission was considering granting Diablo Canyon-1 a low-power license, a set of allegations made by former Diablo Canyon workers caught the attention of one member of an NRC special inspection force, Isa Yin, a senior mechanical engineer and stress analyst with more than ten years' experience in analyzing piping systems.[232] These charges alleged that there had been a widespread quality-assurance breakdown in the seismic design work for major piping in the plant. The rest of the NRC staff acknowledged that there were problems but said they could be resolved after the plant began operation. Yin objected, arguing that inspecting this piping and making changes would be far more difficult after the operation began because of the "intolerable" physical environment in which the inspection would have to be conducted. He was referring to the need to wear bulky anti-contamination clothing, the heat and noise in the reactor containment building, and the difficulty of breathing inside the sealed containment building. Yin said this environment "could discourage additional inspection effort and could hinder any required corrective actions."[233] Nevertheless, the Commission voted to grant the plant a low-power license, with the understanding that Yin's concerns would be resolved before full-power operation was permitted.

Yet, as the full-power review proceeded, Yin found his attempts to carry out his responsibilities thwarted. In July 1984, he resigned from the review process established to investigate the problems he had identified. He asserted that the head of the review team, Richard Vollmer, had prevented him from doing his job. Yin had been given only a few days to review records necessary to resolve his concerns about the Diablo Canyon independent design verification program, and Vollmer rejected his request for additional time. Yin said that "the entire review team only spent two days on work that should have taken a few weeks."[234] He was prevented from reviewing a private firm's independent design verification program records that concluded there were no serious quality-assurance problems. His request to review the new on-site reorganization that was instituted in response to his inspection findings was denied, and he was not allowed to "touch" any follow-ups to whistleblower allegations.[235] Yin told a congressional subcommittee that the NRC review team reports

> contained mostly undocumented reviews and casual observations. There were cases where the inspection sample selected was extremely small, where problems originally identified continued to exist, where review criteria were compromised without technical justification, and where [the NRC] team failed to address the specific program deficiency issues.[236]

In an affidavit filed with the NRC, GAP's legal director, Thomas Devine, said Yin had told him that Diablo Canyon management had "intentionally violated the requirements." According to Devine, Yin described the responses of both the plant management and the NRC staff as "a big Quick Fix."[237] (The "Quick Fix" program at Diablo Canyon involved approximately 16,000 changes to plant equipment. According to Yin, to save time, the changes were made without undertaking the required engineering analysis of their adequacy.[238] The NRC staff acknowledged that this represented a violation of important quality-assurance criteria.[239] PG&E stated that it did the analyses later.[240] In one sense, it is difficult to evaluate the significance of the fact that 16,000 changes were required. Some were certainly important to safety, others less so. However, the fundamental nature of the problem is reflected in the magnitude of the numbers involved: There is no way to be sure that a plant has been built safely when the system for documenting construction quality has gone so greatly awry.) Yin said the staff was "trying to do in a few months what they could not do in two years, and that's asking for trouble. . . . NRR members exhibited their bias by speaking as if their work were done before they had finished their reviews or had heard the licensee's presentation to the staff."[241] Later, Yin said he regretted the harsh tone of the affidavit, but he did not challenge its content.[242]

Yin later told a House subcommittee that the staff review team did not fully address six of the seven conditions that the Commission required to be addressed before it would grant Diablo Canyon a full-power license.[243] "The crux of it is I wasn't allowed to follow up the questions I had. . . ," Yin told a newspaper reporter. He added that "[t]he investigation has never really been carried out."[244]

Moreover, the treatment Yin received from his colleagues on the NRC staff assured that he, and probably many others, would not again put forth differing views. Devine said in his affidavit that, after talking with Yin, "I suggested that Mr. Yin should consider filing a differing professional opinion as a proper channel to express his disagreement. He responded that he had expressed a differing professional opinion in March to the Commission, and it had led to his current situation. He did not want to repeat that."[245]

Vollmer, Yin's supervisor on the review team, recognized the quality-assurance violations but concluded that "although the procedures were not followed, the engineering work that was done was appropriate and *it appeared to the extent that we could tell*, that the items did meet the regulatory criteria."[246] In August 1984 the Commission voted to grant Diablo Canyon Unit 1 a full-power license. Commissioner James Asselstine objected to licensing the plant in the face of outstanding quality-assurance problems:

> With regard to seismic design, the record of this proceeding, allegations filed by former workers at the site and subsequent inspections, including those performed by NRC inspector Isa Yin, all document a widespread quality assurance breakdown in the seismic design work for small bore piping in the plant. This quality assurance breakdown raises some serious questions regarding both the adequacy of quality assurance for other design activities for the plant and the adequacy of the Independent Design Verification Program (IDVP).
>
> When I voted to permit low power operation, it was with the understanding that Mr. Yin and other elements of the NRC staff were in agreement

on the measures needed to resolve those questions prior to a Commission decision authorizing full power operation. . . . Based on the continuing concerns expressed by Mr. Yin regarding the adequacy of the staff's verification efforts and the extent of the seismic design quality assurance breakdown in the case, I am not yet satisfied that the Commission has the information needed to conclude, with a high degree of confidence, that all significant seismic design errors for this plant have been identified and corrected.[247]

Yin stated that the review to resolve his concerns would have taken three to five weeks. After a stay of the Commission decision was issued by the U.S. Court of Appeals (citing in part Asselstine's concerns over the piping problems),[248] Asselstine and Representatives Leon Panetta and Jerry Patterson urged the Commission to take advantage of the time that the plant was to be delayed by establishing a review of the seismic design adequacy of the piping (and to minimize the possibility of future delay).[249] The Commission majority rejected this proposition.

Commissioner Asselstine summed up the implications of the Commission's actions in this case:

I think that there is an unfortunate preoccupation, it has been growing I think over the past year or so, with avoiding licensing delays. That is trying to make licensing decisions on a schedule that the utility says it needs—when it says it is ready to go. . . .

There is nothing wrong with that as long as the Commission makes sure that it only makes its decision if it has the information necessary to conclude that the issuance of a license is justified. But I think part of the problem in this case, and some other cases, has been that there is a growing preoccupation with avoiding licensing delays *per se*, rather than deciding, look, is this plant really ready to go, do we really have the information now to justify the issuance of a license.[250]

The Diablo Canyon case, in both the quality-assurance problems and the earthquake and emergency planning issue described in chapter 4, well illustrates the course the Commission majority appeared to be charting for the future. As Asselstine observed at a congressional hearing on the plant: "The Commission is losing sight of what its regulatory mission really is. Our job is to protect the public health and safety. Our job is not just to issue licenses."[251]

SIX

CONCLUSIONS AND RECOMMENDATIONS

Conclusions

With few exceptions, the performance of the Nuclear Regulatory Commission during its first decade was far from exemplary. The record summarized in this report makes it clear that the agency has not developed into the tough, independent protector of the public health and safety envisioned by the drafters of the Energy Reorganization Act.

The NRC's underlying problem is still one of mindset, a weakness forcefully highlighted by the Kemeny commission in 1979. It is the NRC's indifference and shortsightedness that have allowed so many generic technical problems to persist for so long. The same attitudes have permitted grievous and costly breakdowns in quality control at poorly managed nuclear projects. The evidence in support of these grim conclusions is amply documented in the previous four chapters.

This report has documented how generic safety issues go unresolved year after year and how *resolved* in the NRC's lexicon does not even mean solved in the normal sense of the word but only that the agency has finally decided what the attempted solution should be. Nor does *implemented* imply that any physical changes have been made at the affected reactors; to the NRC, implementation of a particular technical fix means only that utilities have agreed in principle to the desired course of action.

The list of generic safety issues affecting operating plants—only a few of which have been detailed in this report—is long and sobering: ensuring adequate fire protection for reactor safety systems, guaranteeing that key components can function in the harsh environment of a serious accident, detecting and preventing cracks in the major recirculation pipes of boiling water reactors, protecting reactors against the failure to shut down (scram) when necessary, preventing tube deterioration in the steam generators of pressurized water reactors, and many others.

That the NRC classifies problems as "generic" and then fails to deal promptly and decisively with them was a principal criticism of the Kemeny commission in 1979. The criticism is still valid today, and this practice contributes significantly to the economic and safety risks posed by operating reactors and to the *de facto* moratorium on the ordering of new plants.

The NRC's failure to resolve serious technical problems at least partly explains many of the Commission's other shortcomings. If reactors were convincingly safe and free of design and construction defects, there would be little call for internal NRC task forces to ward off needed backfits through cost-benefit calculations of dubious validity. Nor would the agency periodically need to hastily change its regulations (as it did in the equipment qualification issue and is doing with pressurized thermal shock) to cope with plants that do not meet its safety standards.

If the NRC were to resolve the outstanding safety issues and simultaneously demand a high level of competence and integrity from nuclear utilities and other licensees, the licensing process would be far less litigious. The NRC would have little need to perform time-consuming but empty reviews of utility quality-assurance programs, reviews that almost always conclude that the plants are safe despite total breakdowns in these very programs. There would be fewer whistleblowers to discredit, fewer poorly conducted staff investigations, fewer unwanted *sua sponte* actions by licensing boards to quash, fewer misrepresentations by the NRC staff to licensing boards and the Commission itself, hopefully no Department of Justice investigations of the NRC staff,* and less pressure to exclude the public from the licensing process. In short, many (if not most) of the serious problems now afflicting the agency would abate.

Defending the NRC's record to the chairman of a congressional subcommittee, NRC Chairman Nunzio Palladino argued:

> Just as concentrating on the number of times Babe Ruth struck out or on the number of players he left on base can mislead one into believing that he was a poor baseball player, so can one be misled into believing that NRC is not doing a good job if one looks only at its problems. When one looks instead at Babe Ruth's home run record and the number of runs he batted in, one draws a different conclusion about his value as a baseball player. I believe this is also the case with the NRC.[1]

In making this statement, Palladino apparently was claiming that, despite whatever shortcomings others may point to, on the whole the

---

*In late 1984, the NRC revealed that the Department of Justice was investigating possible misconduct by some NRC staff members.

NRC is adequately regulating the nuclear industry and protecting the public. His analogy may be comforting to those who would be satisfied with the current state of affairs, but in our view it serves only as a convenient means of glossing over the clear signs of trouble in the agency.

The fact remains that nuclear power is an inherently dangerous technology requiring the highest standards of care and performance. Soothing reassurances about safety improvements are belied by the serious accidents (TMI and Browns Ferry), the more than 200 "precursors" to core-melt accidents that have occurred in the commercial nuclear program's relatively short history,[2] and the new generic safety problems that are continually surfacing.

The cases we have reviewed in detail are by no means isolated or exceptional. They include many of the difficult problems faced by the Commission in the past decade. The NRC cannot claim success because of its handling of the easy cases; it must be judged by its treatment of the cases that posed the tough issues. Besides being important in and of themselves, they set the standards—clearly observed up and down the ranks of the NRC's staff—for its less-publicized, day-to-day enforcement of the law.

The goal of Congress in establishing the NRC was to alter the institutional dynamics of the regulation of civilian nuclear power by freeing the regulators from the inherent conflict posed by being both regulators and promoters of the technology. The NRC's peformance during its first decade indicates that the approach has not succeeded. The act of dividing the Atomic Energy Commission in two has not had the results hoped for. This outcome was foreshadowed in the new agency's first official act, when it adopted *en masse* the regulations of its predecessor. The great majority of the NRC's senior personnel are likewise holdovers from the AEC.

The Union of Concerned Scientists believes that the record of the first decade demonstrates that the NRC's primary and instinctive allegiance is still to the industry it regulates. Such newer regulatory bodies as the Environmental Protection Agency and the Occupational Safety and Health Administration were born out of a movement to protect the public health and environmental quality; the NRC's historical roots are far different. The AEC's primary task was to bring a civilian nuclear power program into being, to show that the technology invented only for destruction could be adapted to peaceful uses. The regulatory side of that mission was always a stepdaughter. The two most consistent patterns of NRC behavior throughout its first decade illustrate the continuance of that mindset: the agency's inability to disinclination to resolve safety problems and its resolute hostility to public participation. The one period when this pattern was briefly broken was in the aftermath of the accident at Three Mile Island. The agency has since reverted to form—indeed, a strong case can be made that it has regressed as a result of a combination of institutional inertia and the political agenda of the Reagan Administration.

The measures UCS recommends to improve the NRC and enhance the protection of the public are incremental and reflect our view of what is possible in today's political climate. They are based on the assumption that, under the current commissioners, it is not realistic to expect meaningful change in the makeup or institutional instincts of the agency. Under these circumstances, change must come from two other directions: a heightening of public awareness and oversight of the agency's activities, and the establishment of firm nondiscretionary standards in legislation.

Improving the NRC's substantive performance requires a two-pronged approach. First, the agency must be made to seriously come to grips with the underlying technical safety problems. These problems must be solved in the real world, not just in theory or agreement; solutions, when developed, must be incorporated into new plant designs and retrofitted, where needed, into operating plants. Unless that happens, there is little prospect that the commercial nuclear industry will survive, no matter what administrative "reforms" are imposed on the NRC bureaucracy.

Second, the NRC must be made to adopt a much tougher attitude concerning the standards of performance and behavior it expects of licensees. The Commission should be compelled to undertake thorough prelicensing safety reviews; stricter evaluations of the competence, integrity, and financial viability of utility management; aggressive inspections of reactors, both operating and under construction; and more vigorous investigative and enforcement efforts. Unless these steps are taken, more accidents and quality-assurance disasters are inevitable, and the foreclosure of the nuclear option will be all but certain.

Recommendations

Our recommendations to improve the NRC's performance adopt the two-pronged approach just outlined. The first two recommendations relate to the resolution of generic safety problems and would require that safety problems be resolved by clear deadlines and that operating plants be modified accordingly, also on a firm schedule.

(1) Clear deadlines for the resolution of unresolved safety issues should be established by Congress.

(2) The NRC should require the prompt installation of improvements necessary to meet minimum safety requirements at operating plants, regardless of the cost.

The second set of recommendations would set a framework for rational, accountable regulation for the next generation of nuclear plants, should the industry overcome the economic hurdles and the lack of public confidence that now face it. The premise of these recommendations is that important safety advantages can be gained from standardization of nuclear plant design but that, in order for these theoretical advantages to be achieved, the regulatory system must avoid making the same mistakes that have plagued the current generation of plants. In particular, the submitted design must be complete and final and the NRC's review must be thorough, with the resolution of all unresolved safety problems as a condition of approval. In addition, public participation must be encouraged and enhanced. Without these conditions, the risk is great of standardizing unsafe designs that perpetuate past failures. These two recommendations are designed to be applicable to either standardized or individual designs, since the basic principle of thorough review before construction should in any case be a precondition for new plants.

(3) The current process for issuing a construction permit should be modified to allow the issuance of a single combined construction-operating license after the NRC reviews and approves the complete, final design for the plant. The final design should contain the level of detail presently required in an operating license application and must be free of generic defects. A mandatory public hearing would be held, as is now required for a construction permit. No construction at the site would begin until the combined construction-operating license is issued. The NRC would specify unambiguously the design details and construction procedures it approves in the combined license.

(4) Current preoperating license reviews should be modified. Upon completion of the plant's construction, a hearing should

be required to confirm that the plant has been built as designed, that the applicant is capable of safely operating and managing the plant, and that the level of emergency preparedness at and around the plant is adequate. No issues within the scope of the hearing held before issuance of the combined construction-operating license would be reopened unless important new safety questions have arisen since the original permit was granted.

Our third set of recommendations is designed to provide systematic, independent oversight of the activities of the NRC and to enhance the effectiveness of public participation. These recommendations reflect a belief that increasing the accountability of the agency to Congress and the people will improve its performance and a conviction that the agency as presently constituted has failed to firmly enforce its own regulations or aggressively investigate and penalize misconduct.

(5) Congress should establish within the NRC an Office of Inspector General. The inspector general should be appointed by the President and confirmed by the Senate. The inspector general should report directly to Congress and should have the power to investigate allegations, subpoena persons and documents, and refer criminal cases directly to the Justice Department.

(6) Congress should establish an independent, presidentially appointed Nuclear Safety Board, modeled partially after the National Transportation Safety Board, to investigate the causes of accidents and near misses ("precursors") and to review operational experiences that indicate current or potential safety problems. This board should also be directed to "examine on a continuing basis, the performance of the agency and of the nuclear industry in addressing and resolving important public safety issues associated with the construction and operation of nuclear power plants, and in exploring the overall risks of nuclear power. . . . The [board], assisted by its own staff, should report to the President and the Congress at least annually."[3] This recommendation was made by the Kemeny commission.

(7) The NRC should provide financial support for public participation in its proceedings. To ensure that important safety

issues are fully and fairly considered, the NRC should reimburse competent intervenors "who *contribute materially* to rulemaking or licensing efforts by pressing significant concerns that are not being urged by other parties. . . ." Alternatively, the NRC should establish an Office of Public Counsel. Both alternatives were suggested in the Rogovin report by the NRC Special Inquiry Group.[4]

(8) The operations of the Committee to Review Generic Requirements should be recast. Decisions on whether to backfit safety improvements should be made through an open process subject to the Government in the Sunshine Act. Outside experts, from independent groups and from industry, should be invited to address technical issues.

(9) The NRC's inspection and enforcement budget, along with staff levels in its Office of Investigations, should be increased.

We conclude by reaffirming our belief that the NRC's underlying problems start at the top, with the attitudes of the commissioners themselves. If the commissioners demand the resolution of generic issues and a responsive and tough regulatory agency, they will get them. Unfortunately, the converse is also true. Without firm, intelligent, dedicated leadership from the commissioners and the highest levels of the NRC staff, institutional and procedural reforms can be too easily circumvented.

Ultimately, the President and the Senate—through the selection of commissioners—are responsible for the NRC's performance. In this respect, the Union of Concerned Scientists is not optimistic about the future. In the absence of proper presidential leadership, Congress must assume a more assertive oversight role to see that the NRC lives up to its safety-first mandate.

NOTES

1. Introduction

1. Daniel Ford, *Cult of the Atom* (New York: Simon and Schuster, 1982), p.42.

2. Harold Green, quoted in Ford, pp.41–42.

3. Atomic Energy Commission, *Theoretical Possibilities and Consequences of Major Accidents in Large Nuclear Plants*, WASH-740, March 1957.

4. Ford, p.52.

5. 10 CFR Part 50, Appendix A, General Design Criterion 19.

6. Remarks by Sen. Lee Metcalf, *Congressional Record*, 93 Cong. 2 sess., Aug. 15, 1974, p.S15054.

7. Peter Bradford, "Nuclear Hearings, Nuclear Regulation, and Public Safety: A Reflection on the NRC's Indian Point Hearings," speech before the Environmental Defense Fund Associates, New York, Oct. 7, 1982, p.10.

8. For the AEC's response to the ECCS issue, see the following articles in *Science*: "Nuclear Reactor Safety: A Skeleton at the Feast?" May 28, 1971; "Nuclear Reactor Safety: At the AEC the Way of the Dissenter is Hard," May 5, 1972; "Nuclear Safety (I): The Roots of Dissent," Sept. 1, 1972; "Nuclear Safety (II): The Years of Delay," Sept. 8, 1972; "Nuclear Safety (III): Critics Charge Conflicts of Interest," Sept. 15, 1972; "Nuclear Safety (IV): Barriers to Communication," Sept. 22, 1972.

9. Remarks by Sen. Abraham Ribicoff, *Congressional Record*, 93 Cong. 2 sess., Aug. 13, 1974, p.S14869, emphasis added.

10. *Report of the Senate Committee on Government Operations on the Energy Reorganization Act of 1974*, p.58.

11. Ibid., p.83.

12. John Kemeny et al., *Report of the President's Commission on the Accident at Three Mile Island*, October 1979, p.7, original emphasis.

13. Commission meeting transcript, Continuation of Briefing on Action Plan, Dec. 21, 1979, pp.4, 73–74, with attachments.

2. Unresolved "Generic" Safety Problems

1. General Accounting Office, *Management Weaknesses Affect Nuclear Regulatory Commission Efforts to Address Safety Issues Common to Nuclear Power Plants*, GAO/RCED-84-149, Sept. 19, 1984, p.20. (Hereinafter GAO/RCED-84-149.)

2. M. Rowden, "Licensing of Nuclear Power Plants: Reforming the Patchwork Process," *AEI Journal on Government and Society*, January/February 1978, p.46, as cited in *The Nuclear Regulatory Commission*, Staff Report to the President's Commission on the Accident at Three Mile Island, October 1979, p.43. (Hereinafter, Kemeny commission staff report on NRC.)

3. Kemeny commission staff report on NRC, p.43.

4. Gulf States Utilities Co. (River Bend Station, Units 1 and 2), ALAB-444, 6 NRC 760 (1977).

5. Virginia Electric and Power Co. (North Anna Nuclear Power Station, Units 1 and 2), ALAB-491, 8 NRC 245 (1978).

6. Task Action Plan A-3, A-4, A-5, Rev. No. 2, August 1979, emphasis added.

7. *Safety Evaluation Report Related to the Operation of Comanche Peak Steam Electric Station, Units 1 and 2*, NUREG-0797, July 1981, p.C-8, emphasis added.

8. TMI-2 Safety Evaluation Report, Supp. 1, Appendix D, Oct. 22, 1976, as cited in Kemeny commission staff report on NRC, p.44, emphasis added by Kemeny commission staff. See also p.46 of the staff report.

9. John Kemeny et al., *Report of the President's Commission on the Accident at Three Mile Island*, October 1979, p.20. (Hereinafter, Kemeny commission report.)

10. *Investigation of Charges Relating to Nuclear Reactor Safety*, hearings, Joint Committee on Atomic Energy, 94 Cong. 2 sess., Feb. 23, 1976, p.98. (Hereinafter, JCAE hearings on reactor safety.)

11. Technical Safety Activities Report, December 1975, NRC Division of Technical Review, internal working paper, included in JCAE hearings on reactor safety, pp.1202ff.

12. *NRC Program for the Resolution of Generic Issues Related to Nuclear Power Plants*, NUREG-0410, January 1978, pp.iii, 7. (Hereinafter, NUREG-0410.)

13. NUREG-0410.

14. *Identification of Unresolved Safety Issues Relating to Nuclear Power Plants*, NUREG-0510, January 1979. (Hereinafter, NUREG-0510.)

15. NUREG-0410, p.8.

16. H. W. Kendall et al., *The Risks of Nuclear Power Reactors* (Cambridge, Mass: Union of Concerned Scientists, 1977).

17. *Risk Assessment Review Group Report to the U.S. Nuclear Regulatory Commission*, NUREG/CR-0400, September 1978, p.vi.

18. NRC Statement on Risk Assessment and the Reactor Safety Study Report (WASH-1400) in Light of the Risk Assessment Review Group Report, Jan. 18, 1979, p.3.

19. Demetrios Basdekas, NRC, to Robert Budnitz, NRC, Jan. 8, 1979, pp.1, 3. See also James MacKenzie, "Finessing the Risks of Nuclear Power," *Technology Review*, February-March 1984, pp.34–39.

20. Max Carbon, chairman ACRS, letter to NRC Chairman Hendrie, March 21, 1979, p.2.

21. Mitchell Rogovin et al., *three mile island: A Report to the Commissioners and to the Public*, NUREG/CR-1250, Vol. II, Part I, January 1980, p.51.

22. NUREG-0510, p.3.

23. *Identification of New Unresolved Safety Issues Relating to Nuclear Power Plants*, NUREG-0705, March 1981.

24. Ibid., pp.4, 5.

25. *Unresolved Safety Issues Summary*, NUREG-0606, Aug. 17, 1984, and NRC 1983 *Annual Report to the President*, NUREG-1090, June 14, 1984, pp.18–19.

26. Themis Speis, NRC, memorandum to Raymond Fraley, ACRS staff, June 29, 1984.

27. Kemeny commission report, p.65.

28. *A Prioritization of Generic Safety Issues*, NUREG-0933, December 1983. (Hereinafter, NUREG-0933.)

29. GA0/RCED-84-149, p.14.

30. NUREG-0933, p.1.

31. GAO/RCED-84-149, pp.v, 14, 36.

32. Ibid., p.11.

33. *Comparison of Implementation of Selected TMI Action Plan Requirements on Operating Plants Designed by Babcock & Wilcox*, NUREG-1066, May 1984; Licensee Implementation of NUREG-0737 (as of Jan. 1, 1983), undated, FOIA-83-763; Technical Assignment Control System, Multi-Plant Issues Report, Aug. 12, 1983, FOIA-83-703.

34. GAO/RCED-84-149, pp.20–23.

35. Ibid., cover page, pp.23, 28–29.

36. Ibid., pp.20–21, 26–28.

37. Ibid., p.30.

38. Ibid., p.27.

39. Southern California Edison Co. (San Onofre Nuclear Generating Station, Units 2 and 3), CLI-81-33, 14 NRC 1091 (1981).

40. Pacific Gas and Electric Co. (Diablo Canyon Nuclear Power Plant, Units 1 and 2), CLI-84-12, 20 NRC 249 (1984).

41. Metropolitan Edison Co. (Three Mile Island Nuclear Station, Unit No. 1), CLI-84-11, 20 NRC 1, 16 (1984).

42. *NRC Licensing Reform*, hearings, Subcommittee on Energy Conservation and Power, House Committee on Energy and Commerce, 98 Cong. 1 sess., Sept. 23, 1983, pp.104–105; Nunzio Palladino, letter to Sen. George Mitchell, Sept. 29, 1983.

43. *Identification of New Unresolved Safety Issues Relating to Nuclear Power Plants*, NUREG-0705, pp.A-2–A-15.

44. *Unresolved Safety Issues Summary*, NUREG-0606, Nov. 16, 1984.

45. Daniel Ford, Henry Kendall, and Lawrence Tye, *Browns Ferry: The Regulatory Failure* (Cambridge, Mass.: Union of Concerned Scientists, June 10, 1976), p.2. (Hereinafter, Ford et al.)

46. JCAE hearings on reactor safety, March 4, 1976, p.463.

47. Ford et al., pp.4–10.

48. *Browns Ferry Nuclear Plant Fire*, hearings, Joint Committee on Atomic Energy, 94 Cong. 1 sess. 1975, p.188.

49. JCAE hearings on reactor safety, p.466.

50. "Physical Independence of Electric Systems," *Regulatory Guide* 1.75, Revision 2, September 1978.

51. JCAE hearings on reactor safety, pp.964–967, 1085–1090.

52. R. Feit, NRC Division of Reactor Safety Research, memorandum to L. S. Tong, assistant director for water reactor safety research, Aug. 5, 1977, enclosure, p.1.

53. Union of Concerned Scientists, Petition for Emergency and Remedial Action, Nov. 4, 1977.

54. Edson Case, memorandum to the commissioners, Nov. 9, 1977.

55. Petition for Emergency and Remedial Action, CLI-80-21, 11 NRC 707, 716–719 (1980); 10 CFR Part 50, Section 50.48 and App. R.

56. "Fire Protection Rule for Future Plants (SECY-82-267)," SECY-83-269, July 5, 1983.

57. Petition for Emergency and Remedial Action, CLI-78-6, 7 NRC 400 (1978), emphasis added.

58. "NRC Staff Responds to Petition From Union of Concerned Scientists," NRC press release 77-193, Nov. 5, 1977.

59. Petition for Emergency and Remedial Action, CLI-80-21, 11 NRC 707, 711 (1980).

60. NRC Staff Response to "Petition for Extension of Deadline for Compliance with CLI-80-21," July 31, 1981, enclosure, pp.2, 7.

61. "Status of Environmental Qualifications of Electrical Equipment," SECY-82-409, Oct. 8, 1982, enclosure 1, Table 1.

62. 49 Fed. Reg. 45576, "Commissioner Asselstine's Comments on the Notice of Final Rulemaking. . . .," reference omitted.

63. *Union of Concerned Scientists v. Nuclear Regulatory Commission*, 711 F.2d 370, 379 (D.C. Circuit 1983).

64. Brief for Respondent, *Union of Concerned Scientists v. Nuclear Regulatory Commission*, 30, 32. See also Reply Brief for Petitioner, 10–12.

65. Brief for Respondent, *Union of Concerned Scientists v. Nuclear Regulatory Commission*, 20, 29.

66. 49 Fed. Reg. 45572.

67. 49 Fed. Reg. 45576, "Commissioner Asselstine's Comments on the Notice of Final Rulemaking. . . ."

68. "Status Report on Observation of Pipe Cracking at BWRs," SECY-83-267, July 1, 1983. (Hereinafter, SECY-83-267.)

69. "Staff Long Range Plan for Dealing with Stress Corrosion Cracking in BWR Piping," SECY-84-301, July 30, 1984. (Hereinafter, SECY-84-301.)

70. Technical Safety Activities Report, December 1975, NRC Division of Technical Review, internal working paper, included in JCAE hearings on reactor safety, pp.1202ff., item number II.C.A.10.

71. Randell Beck, "Official: NRC puts finances over lives," *The Knoxville Journal*, July 16, 1983; Jane Seegal, UCS, telephone conversation with Richard Vollmer, August 1983.

72. ACRS, letter to Chairman Palladino, Aug. 9, 1983, p.1.

73. Stress Corrosion Cracking in Thick-Wall, Large-Diameter, Stainless Steel Recirculation System Piping at BWR Plants, IE Bulletin No. 82-03, Revision 1, Oct. 28, 1982.

74. SECY-83-267.

75. SECY-84-301.

76. SECY-84-301; Jane Seegal, telephone conversations with Warren Hazelton, NRC, September 1984.

77. SECY-84-301.

78. Seegal, telephone conversation with Hazelton, September 1984.

79. Seegal, telephone conversation with Richard Vollmer, October 1984.

80. "Unprecedented ATWS Fix Urged for Salem in Wake of Recent Incidents," *Nucleonics Week*, March 17, 1983, p.1.

81. *Technical Report on Anticipated Transients Without Scram for Water-Cooled Power Reactors*, WASH-1270, September 1973.

82. *Reactor Safety Study*, WASH-1400 (NUREG-75/014) October 1975, pp.103, 222.

83. *Anticipated Transients Without Scram for Light Water Reactors*, NUREG-0460, Vol.1, April 1978.

84. NUREG-0460, Vol.4, March 1980, p.5.

85. *NRC Fact-Finding Task Force Report on the ATWS Events at Salem Nuclear Generating Station, Unit 1, on February 22 and 25, 1983*, NUREG-0977, pp.xv and 2-1.

86. *Analysis of Loss-of-Feedwater Transients Without Scram For A Westinghouse PWR*, BNL-NUREG-33673, Brookhaven National Laboratory, July 1983.

87. 49 Fed. Reg. 26036.

88. 49 Fed. Reg. 26040.

89. Program Plan on Effectiveness of LWR Regulatory Requirements in Limiting Risk, NRR, Aug. 31, 1984.

90. Chairman Palladino, letter to Sen. George Mitchell, Sept. 29, 1983.

91. 49 Fed. Reg. 16900.

92. Commission meeting transcript, Discussion of Proposed Rule on Backfitting, May 22, 1984, pp.36–37.

93. 49 Fed. Reg. 47038.

94. Office for Analysis and Evaluation of Operational Data (AEOD), *Semiannual Report*, AEOD/S405, January–June 1984, September 1984; 49 Fed. Reg. 47039, "Additional Views of Commissioner Asselstine."

95. 49 Fed. Reg. 47040, "Additional Views of Commissioner Asselstine."

96. "Report of the Task Force on Interim Operation of Indian Point," SECY-80-283, June 12, 1980, p.32.

97. P. Shewmon, chairman, ACRS, letter to Chairman Palladino, June 9, 1982.

98. ACRS Report on the Draft Action Plan for Implementing the Commission's Proposed Safety Goals for Nuclear Power Plants, Sept. 15, 1982, "Additional Comments by Members Myer Bender and Jeremiah J. Ray," pp.6–7.

99. *The Price-Anderson Act—The Third Decade*, NUREG-0957, December 1983, p.I–6; Jane Seegal, telephone conversations with Jack Guerin, GPU Nuclear, 1984 and January 1985.

100. 49 Fed. Reg. 47040, "Additional Views of Commissioner Asselstine."

101. Commission meeting transcript, Discussion of Proposed Rule on Backfitting, May 22, 1984, p.35.

102. Maine Yankee Power Co. (Maine Yankee Atomic Power Station), ALAB-161, 6 AEC 1003, 1007 (1973).

103. Herzel Plaine, memorandum to the Commission, May 4, 1984, pp.4–6. (Hereinafter, Plaine memorandum.)

104. Ibid., p.12.

105. Robert Minogue, memorandum to Commissioner Gilinsky, July 20, 1979.

106. *Power Reactor Development Co. v. International Union*, 367, U.S. 396 (1961).

107. Plaine memorandum, p.9.

108. Commission meeting transcript, Discussion of Proposed Rule on Backfitting, May 22, 1984, pp.29–30.

109. Plaine memorandum, p.3.

110. *American Textile Manufacturers Institute v. Donovan*, 452 U.S. 490 (1981).

111. Charter, Committee to Review Generic Requirements, Revision 1, Jan. 6, 1984.

112. "NRC Chief Evaluates His Tenure," *Energy Daily*, Jan. 9, 1985, p.3.

113. *Nuclear Regulatory Commission Budget Request for Fiscal Years 1984 and 1985*, hearings, Subcommittee on Energy and the Environment, House Committee on Interior and Insular Affairs, 98 Cong. 1 sess., Feb. 22, 1983, p.6.

114. Victor Stello, Jr., memorandum to William Dircks, NRC executive director for operations, April 2, 1982; minutes of CRGR meeting no. 14, May 27, 1982.

115. Michelle Adato, telephone conversation with Daniel Guzy, NRC, Oct. 26, 1984.

116. Stello, memorandum to Dircks, Dec. 10, 1982, p.2.

117. Stello, memoranda to Dircks, Dec. 10, 1982, and Jan. 11, 1983.

118. Stello, memorandum to Dircks, July 24, 1984, enclosure 3.

119. Ibid.

120. Stello, memoranda to Dircks, Nov. 15, 1982; Feb. 8, 1983; May 4, 1983; June 20, 1983.

121. Thomas Murley, NRC, memorandum to Stello, "Summary and Assessment of CRGR Activities to Date," Feb. 28, 1983.

122. *TMI-2 Lessons Learned Task Force Final Report*, NUREG-0585, October 1979, pp.2–7.

123. "NUREG-1070, 'NRC Policy on Future Reactor Designs: Decisions on Severe Accident Issues in Nuclear Power Plant Regulations,'" SECY-84-370, Sept. 19, 1984, enclosure 1, pp.12–13, Aug. 23, 1984. (Hereinafter, SECY-84-370.)

124. Commission meeting transcript, Discussion on Severe Accident Program for Nuclear Power Reactors -Revised Policy Statement, Oct. 9, 1984, p.61. (Hereinafter, policy statement transcript.)

125. *Inside NRC*, Oct. 15, 1984, p.10.

126. Policy statement transcript, p.7.

127. Ibid., p.5.

128. Proposed Policy Statement on Safety Goals for Nuclear Power Plants, Feb. 11, 1982.

129. William Dircks, memorandum to Harold Denton, Robert Minogue, and Richard DeYoung, Dec. 17, 1982.

130. Stello, memorandum to Dircks, May 8, 1984, pp.3–4.

131. Ibid., pp.4–5.

132. SECY-84-370, enclosure 1, p.14.

3. The Public as Adversary

1. John Kemeny et al., *Report of the President's Commission on the Accident at Three Mile Island*, October 1979, p.20.

2. *NRC Licensing Reform*, hearing, Subcommittee on Energy Conservation and Power, House Committee on Energy and Commerce, 98 Cong. 1 sess., Sept. 23, 1983, pp.104–105; NRC Chairman Palladino, letter to Sen. George Mitchell, Sept. 29, 1983.

3. Peter Bradford, statement, Subcommittee on Environment, Energy and Natural Resources, House Committee on Government Operations, 96 Cong. 2 sess., July 2, 1980.

4. "Further Commission Guidance for Power Reactor Operating Licenses," Revised Statement of Policy, Dec. 18, 1980.

5. Peter Bradford, testimony, *Nuclear Licensing and Regulatory Reform*, Subcommittee on Nuclear Regulation, Senate Committee on Environment and Public Works, 98 Cong. 1 sess., July 14, 1983, p.562.

6. 10 CFR 2.764(f). Also see 10 CFR 2.764 (f) (2).

7. Mitchell Rogovin et al., *three mile island: A Report to the Commissioners and to the Public*, NUREG/CR-1250, January 1980, p.139. (Hereinafter, Rogovin report.)

8. 10 CFR 2.714(b).

9. Southern California Edison Co. et al. (San Onofre Nuclear Generating Station, Units 2 and 3) LBP-83-47, 18 NRC 228.

10. Metropolitan Edison Co. (Three Mile Island Station, Unit No. 1) LBP-81-59, 14 NRC 1211, 1401 (1981).

11. Ibid., pp.1401–1404.

12. John Stolz, NRC, letter to Henry Hukill, GPU Nuclear, April 25, 1984, enclosure.

13. Public Service Co. of New Hampshire et al. (Seabrook Station, Units 1 and 2) ALAB-390, 5 NRC 733 (1977).

14. 45 Fed. Reg. 55402, 55403.

15. Director's Decision, 14 NRC 279, 286 (1981).

16. *Belotti v. NRC*, 725 F.2d 1380 (D.C. Cir., 1984).

17. Texas Utilities Generating Co. et al. (Comanche Peak Steam Electric Station, Units 1 and 2) LBP-81-23, 14 NRC 159 (1981).

18. *Scenic Hudson Preservation Conference v. Federal Power Commission*, 354 F.2d 608, 614, 620 (2nd Cir. 1965).

19. 10 CFR 2.760(a).

20. Southern California Edison Co. et al. (San Onofre Nuclear Generating Station, Units 2 and 3), LBP-81-36, 14 NRC 691 (1981).

21. Southern California Edison Co. et al. (San Onofre Nuclear Generating Station, Units 2 and 3), CLI-81-33, 14 NRC 1091 (1981), "Separate Views of Commissioner Gilinsky Regarding the San Onofre *Sua Sponte* Issue."

22. Southern California Edison Co. et al. (San Onofre Nuclear Generating Station, Units 2 and 3), CLI-81-33, 14 NRC 1091 (1981), "Separate Views of Commissioner Bradford Regarding the San Onofre *Sua Sponte* Issue," footnotes omitted.

23. Metropolitan Edison Co. (Three Mile Island Station, Unit No. 1) CLI-80-16, 11 NRC 674 (1980).

24. Northern Indiana Public Service Co. (Bailly Generating Station, Nuclear-1) CLI-79-11, 10 NRC 733 (1979).

25. *Illinois v. NRC*, No. 80-1163 (D.C. Cir., July 1, 1981).

26. Houston Lighting and Power Co. et al. (South Texas Project, Units 1 and 22), CLI-81-28, 14 NRC 933 (1981).

27. Pacific Gas and Electric Co. (Diablo Canyon Nuclear Power Plant, Units 1 and 2) CLI-81-22, 14 NRC 598 (1981).

28. Pacific Gas and Electric Co. (Diablo Canyon Nuclear Power Plant, Units 1 and 2), "Dissenting Views of Commissioner Asselstine," CLI-84-12, Aug. 10, 1984.

29. 49 Fed. Reg. 49640.

30. Texas Utilities Generating Co. et al. (Comanche Peak Steam Electric Station, Units 1 and 2) LBP-81-38, 14 NRC 767 (1981).

31. Texas Utilities Generating Co. et al. (Comanche Peak Steam Electric Station, Units 1 and 2) CLI-81-36, 14 NRC 1111, 1115, "Separate Views of Commissioner Bradford" (1981).

32. Cincinnati Gas and Electric Co. et al. (Zimmer Nuclear Power Station, Unit 1) LBP-82-54, 16 NRC 210, 214 (1982).

33. NRC Staff Response to Miami Valley Power Project Motion for Leave to File Contentions, Docket 50-358, June 11, 1982, pp.4–5.

34. Cincinnati Gas and Electric Co. et al. (Zimmer Nuclear Power Station, Unit 1) LBP-82-54, 16 NRC 214–215 (1982).

35. Cincinnati Gas and Electric Co. et al. (Zimmer Nuclear Power Station, Unit 1) CLI-82-20, 16 NRC 109 (1982).

36. Ibid., "Dissenting Opinion of Commissioner Asselstine," 16 NRC 109, 115, original emphasis.

37. Consolidated Edison Co. of New York (Indian Point, Unit 2), Power Authority of the State of New York (Indian Point, Unit 3) CLI-82-15, 16 NRC 27 (1982); letter to the ASLB from Samuel Chilk, secretary, Aug. 23, 1982 (later formalized as CLI-82-25, Sept. 20, 1982).

38. "Indian Point Judge Criticizes NRC," *Newsday*, Sept. 12, 1982, p.7.

39. Commission Memorandum, Consolidated Edison Co. of New York (Indian Point, Unit 2), Power Authority of the State of New York (Indian Point, Unit 3), Aug. 20, 1982.

40. Separate Views of Commissioner Gilinsky, Commission Memorandum, Consolidated Edison Co. of New York (Indian Point, Unit 2), Power Authority of the State of New York (Indian Point, Unit 3), Aug. 20, 1982.

41. Louis Carter, testimony, *Indian Point and NRC Safety Procedures*, hearings, Subcommittee on Energy, Conservation, and Power, House Committee on Energy and Commerce, 97 Cong. 2 sess., Sept. 24, 1982, p.190.

42. "Indian Point Judge Criticizes NRC," *Newsday*, Sept. 12, 1982, p.7.

43. Consolidated Edison Co. of New York (Indian Point, Unit 2), Power Authority of the State of New York (Indian Point, Unit 3) LBP-82-61, 16 NRC 560 (1982).

44. Carter, letter to the commissioners, Sept. 1, 1982.

45. Peter Bradford, "Nuclear Hearings, Nuclear Regulation, and Public Safety: A Reflection on the NRC's Indian Point Hearings," speech before the Environmental Defense Fund Associates, New York, Oct. 7, 1982, pp.13–14.

46. 292d Meeting of the ACRS, transcript, Aug. 10, 1984, p.510. (Hereinafter, ACRS meeting transcript.)

47. Gulf States Utility Co. (River Bend Units 1 and 2), ALAB-183, RAI-74-3, March 12, 1974, S1. Op., pp.10–12.

48. B. Paul Cotter, Jr., memorandum to Commissioner Ahearne, May 1, 1981, p.8.

49. Rogovin report, Vol.1, pp.143–144, original emphasis.

50. ACRS meeting transcript, pp.509–510.

51. Attachment to ACRS meeting transcript.

52. Boston Edison Co. et al. (Pilgrim Nuclear Power Station, Unit 2) LBP-81-3, 13 NCR 103, 194–197 (1981).

53. *Metropolitan Siting—A Historical Perspective*, NUREG-0478, October 1978, pp.7–8, 11–12.

54. 10 CFR Part 50, section 50.48 and Appendix R; Petition for Emergency and Remedial Action, CLI-80-21, 11 NRC 707, 716–19 (1980).

55. 10 CFR 50.49; Petition for Emergency and Remedial Action, CLI-80-21, 11 NRC 707, 710–16 (1980).

56. Docket 50-275: Intervenors Motion to Add Quality Assurance Contention, April 29, 1977; Request for Directed Certification, Dec. 3, 1980; Order Suspending License, CLI-81-30, 14 NRC 950 (1981).

57. *Licensing Process at Diablo Canyon Nuclear Powerplant*, hearing, Subcommittee on Energy and the Environment, House Committee on Interior and Insular Affairs, 98 Cong. 2 sess., Aug. 30, 1984, p.16.

58. *Emergency Preparedness, Emergency Response*, Staff Report to the President's Commission on the Accident at Three Mile Island, October 1979, pp.14–15.

59. *Report of Emergency Preparedness and Response Task Force*, Staff Report to the President's Commission on the Accident at Three Mile Island, October 1979, pp.22–31, 52–65.

60. Docket 50-443: SAPL Findings of Fact and Conclusions of Law, Dec. 12, 1975; Motion to Reopen Record Filed on Behalf of Intervenors SAPL and Audubon on Issues of Financial Qualifications, Need for Power and Cost Benefit Analysis, Dec. 30, 1975; Renewed Motion to Reopen Hearing Filed on Behalf

of Intervenor SAPL on Issue of Financial Qualifications, Feb. 13, 1976; Motion to Reopen on Financial Qualifications, June 2, 1977; Motion to Reopen Record on Financial Qualification, Nov. 30, 1978; Request for an Order to Show Cause Why Construction Permits for the Proposed Nuclear Power Plants at Seabrook Should Not Be Suspended or Revoked, Feb. 5, 1982.

61. Suffolk County Executive Peter Cohalan, testimony, Subcommittee on Oversight and Investigations, House Committee on Interior and Insular Affairs, Exhibit 3, 98 Cong. 1 sess., April 18, 1983.

62. Rep. Edward Markey, statement before the Energy and Environment Subcommittee, House Committee on Interior and Insular Affairs, 98 Cong. 2 sess., July 24, 1984.

63. Markey, letter to Chairman Palladino, March 13, 1984.

64. Peter Bradford, testimony, *Nuclear Licensing and Regulatory Reform*, hearing, Subcommittee on Nuclear Regulation, Senate Committee on Environment and Public Works, 98 Cong. 1 sess., July 14, 1983, p.561.

65. *Energy and Water Development Appropriations for 1981*, hearing, Subcommittee on Energy and Water Development, House Committee on Appropriations, 96 Cong. 2 sess., April 17, 1980, pp.267–268, 311.

66. Office of Nuclear Reactor Regulation, Monthly Status Report to Congress (the Bevill report), Jan. 30, 1981.

67. *Nucleonics Week*, Feb. 19, 1981, p.7.

68. William Lee, testimony, *Nuclear Powerplant Licensing Delays and the Impact of the Sholly Versus NRC Decision*, hearing, Subcommittee on Nuclear Regulation, Senate Committee on Environment and Public Works, 97 Cong. 1 sess., March 25, 1981, p.50. (Hereinafter, hearing on licensing delays.)

69. Ellyn Weiss, testimony, hearing on licensing delays, p.95.

70. *Licensing Speedup, Safety Delay: NRC Oversight*, House Committee on Government Operations, H. Rept. 97–277, 97 Cong. 1 sess., Oct. 20, 1981, pp.37–39, 42–43.

71. Long Island Lighting Co. (Shoreham Nuclear Power Plant, Unit 1), "Separate Views of Commissioner Gilinsky," CLI-84-8, May 16, 1984.

72. 5 U.S.C.557 (d) (1) (A)-(B) and 10 CFR 2.780.

73. Chairman Palladino, memorandum to the commissioners, March 20, 1984.

74. Suffolk County and State of New York Request for Recusal and, Alternatively, Motion for Disqualification of Chairman Palladino, Docket 50-322, June 5, 1984, p.19.

75. Long Island Lighting Co. (Shoreham Nuclear Power Plant, Unit 1), Memorandum of Chairman Palladino, Sept. 24, 1984, p.22.

76. Ibid., p.23.

77. Long Island Lighting Co. (Shoreham Nuclear Power Plant, Unit 1), Memorandum and Order Scheduling Hearing on LILCO's Supplemental Motion for Low-Power Operating License, April 6, 1984.

78. Long Island Lighting Co., Supplemental Motion for Low Power Operating License, Docket 50-322, March 20, 1984, p.19.

79. Rep. Edward Markey, letters to Chairman Palladino, March 28, 1984; April 12, 1984; April 24, 1984; May 10, 1984.

80. Palladino, letter to Markey, April 5, 1984.

81. Commission meeting transcript, Discussion of Possible Steps to Avoid Licensing Delays, April 24, 1984, pp.5–6.

82. Ibid., p.6.

83. *Mario M. Cuomo et al. v. U.S. Nuclear Regulatory Commission*, 84–1264, April 25, 1984.

84. Rep. Morris Udall, opening statement, *Licensing Process at Shoreham Nuclear Powerplant*, hearing, Subcommittee on Energy and the Environment, House Interior Committee, 98 Cong. 2 sess., May 17, 1984, p.2.

85. Long Island Lighting Co. (Shoreham Nuclear Power Plant, Unit 1), "Separate Views of Commissioner Gilinsky," CLI-84-8, 19 NRC 1154, 1159 (1984).

86. Long Island Lighting Co. (Shoreham Nuclear Power Plant, Unit 1), "Additional Views of Commissioner Asselstine," CLI-84-8, 19 NRC 1154 (1984).

87. Suffolk County and State of New York Request for Recusal and, Alternatively, Motion for Disqualification of Chairman Palladino, Docket 50-322, June 6, 1984.

88. Long Island Lighting Co. (Shoreham Nuclear Power Plant, Unit 1), Memorandum of Chairman Palladino, Sept. 24, 1984.

89. Long Island Lighting Co. (Shoreham Nuclear Power Plant, Unit 1) LBP-83-71, 17 NRC 593, 601 (1983).

90. Long Island Lighting Co. (Shoreham Nuclear Power Plant, Unit 1), "Separate Views of Commissioner Gilinsky: Shoreham Emergency Planning Lower Power Operations," CLI-83-17, 17 NRC 1032, 1035 (1983).

91. "State and Lilco Holding Talks on Rescue Plan," *New York Times*, Feb. 19, 1984, p.1; "Lilco Chief Sees 4 Months Left to Save Utility," *New York Times*, Feb. 22, 1984, p.B2.

92. Commissioner Gilinsky, memorandum to Chairman Palladino, May 29, 1984.

93. Ibid., added note of Commissioner Asselstine, May 31, 1984.

94. Commission meeting transcript, Continuation of 4/24 Discussion on Possible Steps to Avoid Licensing Delays to Include Discussion of Last Minute Allegations, June 15, 1984, p.51.

95. Ibid., p. 42.

96. William Dircks, memorandum to Chairman Palladino, June 25, 1984.

97. 5 U.S.C. s. 556 (d).

98. S. 893, s. 102 (98 Cong. 1 sess.); H.R. 2512, s. 102 (98 Cong. 1 sess.).

99. Commissioner Gilinsky, statement, *Nuclear Licensing Reform*, hearings, Subcommittee on Energy and the Environment, House Committee on Interior and Insular Affairs, 98 Cong. 1 sess., June 9, 1983, p.138.

100. Peter Bradford, testimony, *Nuclear Licensing and Regulatory Reform*, hearing, Subcommittee on Nuclear Regulation, Senate Committee on Environment and Public Works, 98 Cong. 1 sess., July 14, 1983, p.560.

101. Ibid., pp.567–568.

102. Commission meeting transcript, Discussion of RRTF—Admin. Proposals—Rev. to Part 2, Nov. 16, 1983, pp.62–63.

103. 49 Fed. Reg. 14698.

104. David Welborn and William Lyons, University of Tennessee-Knoxville, and Larry Thomas, University of Baltimore, *Implementation and Effects of the Federal Government in the Sunshine Act* (Washington, D.C.: Administrative Conference of the United States, June 1984), pp.4–5. This report represents only the views of the authors, not necessarily those of the conference.

105. Ibid., pp.59–67.

106. Ibid., p.61.

107. Ibid., pp.77–78.

108. Ibid., pp.37–38.
109. *Common Cause et al. v. Nuclear Regulatory Commission*, 674 F. 2d 921 (D.C. Circuit, Feb. 26, 1982).
110. *Philadelphia Newspapers Inc. v. Nuclear Regulatory Commission*, 727 F.2d 1195 (Feb. 10, 1984).

4. Arbitrary Enforcement of Regulations

1. Vermont Yankee Nuclear Power Corp. (Vermont Yankee Nuclear Power Station) ALAB-138, RAI-73-7, 520, 528 (1973), emphasis added.
2. *Union of Concerned Scientists v. Nuclear Regulatory Commission*, 711 F.2d 370 (D.C. Cir. 1983); *New England Coalition on Nuclear Pollution v. Nuclear Regulatory Commission*, 727 F.2d 1127, (D.C. Cir. 1984); *Mario M. Cuomo v. Nuclear Regulatory Commission*, 84-1264, April 25, 1984; *Union of Concerned Scientists v. Nuclear Regulatory Commission*, 735 F.2d 1437 (D.C. Cir. 1984), amended July 2, 1984; *San Luis Obispo Mothers for Peace v. Nuclear Regulatory Commission*, 84-1410, Aug. 17, 1984.
3. 10 CFR 50.54(s) (2) (ii).
4. Boyce Grier, administrator, Region I, NRC, letters to J. D. O'Toole, Consolidated Edison Co., and G.T. Berry, Power Authority of the State of New York, April 24, 1981.
5. R. Jaske, FEMA, memorandum to Director, Division of Emergency Preparedness, Office of Inspection and Enforcement, NRC, Aug. 19, 1981; Consolidated Edison Co. of New York (Indian Point, Unit No. 2), Power Authority of the State of New York (Indian Point, Unit No. 3), CLI-82-38, 16 NRC 1698 (1982).
6. Grier, letter to Con Ed and PASNY, Aug. 24, 1981.
7. Frank Petrone, Region II, FEMA, memorandum for L. M. Thomas, State and Local Programs and Support, FEMA, July 29, 1982.
8. Lee Thomas, FEMA, letter to William Dircks, NRC, Dec. 17, 1982.
9. Resolution #320, Rockland County Legislature, May 18, 1982.
10. Dave McLoughlin, FEMA, letter to Dircks, NRC, April 15, 1983.
11. Consolidated Edison Co. of New York (Indian Point, Unit No. 2), Power Authority of the State of New York (Indian Point, Unit No. 3), CLI-83-11, 17 NRC 731 (1983).
12. J. S. Bragg, FEMA, letter to Chairman Palladino, NRC, June 8, 1983.
13. Consolidated Edison Co. of New York (Indian Point, Unit No. 2), Power Authority of the State of New York (Indian Point, Unit No. 3), CLI-83-16, 17 NRC 1006 (1983).
14. Ibid., p.1014.
15. Ibid., "Dissenting Opinion of Commissioner Asselstine," p.1031.
16. Ibid., "Separate Views of Commissioner Gilinsky Regarding Indian Point," p.1028.
17. Fracture Toughness Requirements, 10 CFR 50, Appendix G, Section IV.B.
18. Draft NRC Staff Evaluation of Pressurized Thermal Shock, Sept. 13, 1982, Table P-1.
19. 49 Fed. Reg. 4501.
20. 49 Fed. Reg. 4500.
21. Draft NRC Staff Evaluation of Pressurized Thermal Shock, Sept. 13, 1982.
22. 49 Fed. Reg. 4498.

23. 49 Fed. Reg. 4500.

24. *Investigation of Charges Relating to Nuclear Reactor Safety*, hearings, Joint Committee on Atomic Energy, 94 Cong. 2 sess., Feb. 18, 23, and 24 and March 2 and 4, 1976, p.1192.

25. 49 Fed. Reg. 4500, emphasis added.

26. Ibid.

27. Formerly 10 CFR 55.25.

28. "Issuance of NUREG-0094, NRC Licensing Guide Entitled, 'A Guide for the Licensing of Facility Operators, Including Senior Operators,' " SECY-76-341, June 28, 1976, enclosure 1.

29. 49 Fed. Reg. 31702, "Separate Views of Former Commissioner Gilinsky on Proposed Amendments to 10 CFR Part 55"; Commission meeting transcript, Interpretation of 10 CFR 55.25(b) on Reactor Operator Examinations, April 13, 1984, p.25. (Hereinafter, interpretation transcript.)

30. James Fitzgerald, memorandum for commissioner Gilinsky, "Licensed Reactor Operator Examinations," April 6, 1984.

31. Herzel Plaine, memorandum for the commissioners, April 12, 1984.

32. "Licensed Reactor Operator Examinations," SECY-84-152, April 11, 1984.

33. Interpretation transcript, pp.32, 50.

34. Ibid., pp.4–5.

35. 49 Fed. Reg. 31701, "Separate Views of Former Commissioner Gilinsky on Proposed Amendments to 10 CFR Part 55."

36. Ibid.

37. John Kemeny et al., *Report of the President's Commission on the Accident at Three Mile Island*, October 1979, p.50.

38. Interpretation transcript, p.48.

39. 49 Fed. Reg. 31701, "Separate Views of Former Commissioner Gilinsky on Proposed Amendments to 10 CFR Part 55."

40. Interpretation transcript, p.61.

41. NRC staff meeting transcript, A Meeting on TDI Diesel Generators, Jan. 26, 1984, attachments.

42. Ibid., p.8.

43. Ibid., pp.95–96.

44. NRC Staff's Response to Suffolk County's Motion to Admit Supplemental Diesel Generator Contentions, Docket No. 50-322, Feb. 14, 1984, p.12 and note 7.

45. Commission meeting transcript, Discussion of Shoreham Licensing Proceeding, April 23, 1984, p.59.

46. NRC Staff Response to LILCO's Supplemental Motion for Low Power Operating License, Docket No. 50-322, March 30, 1984, p.4.

47. Long Island Lighting Co. (Shoreham Nuclear Power Plant, Unit 1), "Separate Views of Commissioner Gilinsky," CLI-84-8, 19 NRC 1154, 1159 (1984).

48. Commission meeting transcript, Discussion of Shoreham Licensing Proceeding, April 23, 1984.

49. Herzel Plaine, memorandum to the Commission, April 2, 1984.

50. Commission meeting transcript, Oral Argument on Shoreham, May 7, 1984, pp.57–58.

51. Ibid., p.62.

52. Long Island Lighting Co. (Shoreham Nuclear Power Plant, Unit 1) CLI-84-8, 19 NRC 1154 (1984).

53. Commissioner Gilinsky, letter to Rep. Morris Udall, May 29, 1984.

54. NRC staff, Order Requiring Diesel Generator Inspection (Effective Immediately), Docket No. 50-416, May 22, 1984.

55. Commission meeting transcript, Discussion of Grand Gulf Order, June 1, 1984, p.18.

56. NRC staff, Order Requiring Diesel Generator Inspection (Effective Immediately), Docket No. 50-416, May 22, 1984.

57. Herzel Plaine, memorandum to the Commission, May 30, 1984.

58. Rep. Richard Ottinger, opening statement, hearing, Subcommittee on Energy Conservation and Power, House Committee on Energy and Commerce, 98 Cong. 2 sess., June 21, 1984.

59. Commission meeting transcript, Discussion of Grand Gulf Order, June 1, 1984, p.23.

60. Ibid., p.48.

61. Ibid, p.27.

62. Ibid., p.40.

63. Long Island Lighting Co. (Shoreham Nuclear Power Plant, Unit 1) CLI-84-8, 19 NRC 1154 (1984).

64. "Needs and Standards for Exemptions," SECY 84-290, July 17, 1984.

65. Commission meeting transcript, Discussion of Commission Practice in Granting Exemptions, July 25, 1984, p.4.

66. Ibid., pp.26–28, emphasis added.

67. Ibid., pp.29–30.

68. Ibid., pp.39–40.

69. Ibid., p.46.

70. Ibid., p.70.

71. Ibid., p.85.

72. Ibid., pp.86–87.

73. "Proposed Amendments to 10 CFR 50.12, 'Specific Exemptions,'" SECY-84-356, Sept. 10, 1984.

74. Commission meeting transcript, Discussion of Commission Practice in Granting Exemptions, July 25, 1984, pp.96–97.

75. Ibid., p.98.

76. *Union of Concerned Scientists v. Nuclear Regulatory Commission*, 735 F.2d 1437, 1438, May 25, 1984.

77. Pacific Gas and Electric Co. (Diablo Canyon Nuclear Power Plant, Units 1 and 2), "Dissenting Views of Commissioner Asselstine," CLI-84-12, 20 NRC 249, 259 (1984). (Hereinafter, Asselstine's dissent.)

78. Asselstine's dissent, p.6.

79. Asselstine's dissent, pp.7–9.

80. William Dircks, memorandum to the commissioners, June 22, 1982, p.1, as cited in Asselstine's dissent, p.3.

81. Dircks, memorandum to Chairman Palladino, Jan. 13, 1984, p.2., footnote 2, as cited in Asselstine's dissent, p.4.

82. NRC staff, memorandum to the Commission, June 22, 1982, enclosure, pp.3–4, as cited in Asselstine's dissent, pp.3–4.

83. NRC Staff's Memorandum Regarding Consideration of Effects of Earthquakes on Emergency Planning (CLI-84-8), May 3, 1984, p.3, footnote 2, as cited in Asselstine's dissent, p.5.

84. Asselstine's dissent, p.5., footnote 1.

85. Rep. Richard Ottinger, letter to Chairman Palladino, Oct. 26, 1984.

86. Ibid.

87. Ibid.

88. *San Luis Obispo Mothers for Peace v. Nuclear Regulatory Commission*, D.C. Cir., No. 84-1410, Aug. 17, 1984.

89. Ottinger, letter to Attorney General William French Smith, Nov. 1, 1984.

90. Ottinger, letter to Palladino, Oct. 26, 1984.

91. Ibid.

92. Palladino, letter to Rep. Edward Markey, Oct. 29, 1984.

93. Ibid., "Additional Comments of Commissioner Asselstine," attached.

94. Commission meeting transcript, Discussion of Earthquake and Emergency Planning for Diablo Canyon, July 25, 1984, p.3.

95. Ibid. p.11.

96. Commission meeting transcript, Discussion of Earthquakes and Emergency Planning for Diablo Canyon and Discussion of Stay Motion, July 30, 1984, p.61.

97. Commission meeting transcript, Discussion of Earthquake and Emergency Planning for Diablo Canyon, July 25, 1984, p.54.

98. 42 U.S.C. 2239 Sec. 189(a).

99. Metropolitan Edison Co. (Three Mile Island Nuclear Station, Unit No. 1) CLI-80-26, 11 NRC 789 (1980).

100. *Sholly v. Nuclear Regulatory Commission*, 651 F.2nd 780 (D.C. Cir. 1980).

101. Chairman Hendrie, hearings, *Nuclear Regulatory Commission Operating Licensing Process*, Subcommittee on Energy and the Environment, House Committee on Interior and Insular Affairs, 97 Cong. 1 sess., April 9, 1981, pp.8, 11, 29.

102. 42 U.S.C. 2239 Sec. 189a(2).

103. H. Rept. 97-884, 97 Cong. 2 sess., p.37 (1982).

104. Ibid.

105. Ibid., emphasis added.

106. 10 CFR 50.92(c) (1-3).

107. 48 Fed. Reg. 14872.

108. H. Rept. 97-884, 97 Cong. 2 sess., p.37.

109. Darrell Eisenhut, memorandum to Harold Denton, Sept. 18, 1984.

110. 48 Fed. Reg. 26930, 52805, 52808, 52811.

111. Eisenhut, letter to Henry Hukill, GPU Nuclear, Aug. 23, 1982.

112. Ibid. See also *NRC Authorization for F.Y. 1984-5*, hearings, Subcommittee on Energy Conservation and Power, House Committee on Energy and Commerce, 98 Cong. 1 sess., June 8, 1983, p.528. (Hereinafter, authorization hearings.)

113. *Safety Evaluation Report Related to Steam Generator Tube Repair and Return to Operation*, NUREG-1019, November 1983, p.2.

114. 48 Fed. Reg. 24231.

115. 48 Fed. Reg. 14870.

116. 48 Fed. Reg. 24231.

117. Authorization hearings, pp.529, 531.

118. Ibid., p.531.

119. Views of Commissioner James K. Asselstine on the NRC Staff's NSHC Determination on the Three Mile Island Unit 1 (TMI-1) License Amendment Application for Steam Generator Repairs, January 1984, unpublished, p.7.

120. *Inside NRC*, April 2, 1984, p.14.

121. Commission meeting transcript, Discussion/Possible Vote on TMI Steam Generators, Jan. 10, 1984.

122. *Inside NRC*, April 2, 1984, p.14.

123. 48 Fed. Reg. 1486; "Study on Significant Hazards," SECY-83-337, Aug. 15, 1983, p.2.

124. 127 *Congressional Record* H.8156, 97 Cong. 1 sess.

125. S. Rept. 97-113, 97 Cong. 1 sess., p.15.

126. Herzel Plaine and Guy Cunningham, memorandum to the Commission, March 1983.

127. 48 Fed. Reg. 14869.

128. Comments of the Union of Concerned Scientists, May 6, 1983, pp.12–14; Comments on Interim Final Rule Regarding No Significant Hazards Considerations, State of Maine, Department of the Attorney General, May 6, 1983, pp.6–9.

129. "Study on Significant Hazards," SECY-83-337, Aug. 15, 1983.

130. "Oconee Unit No. 3—Spent Fuel Pool Expansion," SECY-83-334, Aug. 15, 1983.

131. Atomic Energy Act, section 189a(2) (c) (ii); 10 CFR 50.91(a).

132. 48 Fed. Reg. 36355, 50637, 46909.

133. 48 Fed. Reg. 31929, 44952, 49604.

134. Herzel Plaine, memorandum to commissioners, May 30, 1984.

135. "Additional Views of Commissioner Asselstine," attached to Chairman Palladino, letter to Rep. Edward Markey, July 23, 1984; Oversight Hearing on the Grand Gulf Nuclear Powerplant, unbound transcript, Subcommittee on Energy and the Environment, House Committee on Interior and Insular Affairs, 98 Cong. 2 sess., July 24, 1984.

136. 48 Fed. Reg. 33083, emphasis added.

137. Draft of Request for Publication in Monthly FR Notice—Notice of Consideration of Issuance of Amendment of Facility Operating License and Proposed No Significant Hazards Consideration Determination and Opportunity for a Hearing. Final notice published June 27, 1983.

138. Marjorie Rothschild, NRC, memorandum to Robert Clark, NRC, June 13, 1983, FOIA 84-84.

139. 42 U.S.C. Section 303 (a).

140. Rep. Richard Ottinger, letter to Chairman Palladino, April 9, 1982.

141. Ibid.

142. Ibid.

143. Palladino, letter to Ottinger, March 1, 1982, as cited in Ottinger, letter to Palladino, April 9, 1982.

144. "Added Views of Commissioner Bernthal," attached to Palladino letter to Rep. Edward Markey, Jan. 25, 1984.

145. Richard Udell, Subcommittee on Oversight and Investigations, House Committee on Interior and Insular Affairs, telephone conversation with Michelle Adato, UCS, Jan. 15, 1985.

146. Stanley Scoville, House Committee on Interior and Insular Affairs, telephone conversation with Jane Seegal, UCS, January 1985.

147. Palladino, letter to Markey, April 26, 1984, emphasis added.

148. Commission meeting transcript, Briefing on Markey Letter, April 24, 1984.

149. *Licensing Process at Shoreham Nuclear Powerplant*, hearing, Subcommittee on Energy and the Environment, House Committee on Interior and Insular Affairs, 98 Cong. 2 sess., May 17, 1984, p.38.

150. Commissioner Ahearne, letter to Rep. Morris Udall, January 1980.

151. William Dircks, memorandum to Chairman Ahearne and Commissioners Gilinsky, Hendrie, and Bradford, Feb. 17, 1981.

152. Rep. Morris Udall and Rep. Manuel Lujan, Jr., letter to Commissioner Hendrie, NRC, March 4, 1981.

153. Udall, hearings, *NRC Budget Request for FY 84 and 85*, Subcommittee on Energy and the Environment, House Committee on Interior and Insular Affairs, 98 Cong. 1 sess., Feb. 22, 1983, p.22.

5. Self-Regulation and Nuclear Camaraderie

1. Jonathan Lash, Katherine Gillman, and David Sheridan, A Season of Spoils (New York: Pantheon), pp.32–36.

2. John Zerbe, memorandum to the commissioners, Aug. 7, 1984; "Indiana Utility Scraps $7.7 Billion Marble Hill Because It Costs Too Much," *Nucleonics Week*, Jan. 19, 1984, p.1; "Michigan Utility Says It Will Close Midland Project," *The Wall Street Journal*, July 17, 1984, p.2; "Shutdown at Midland Authorized," *The New York Times*, July 17, 1984.

3. *Nuclear Licensing Reform*, hearing, Subcommittee on Energy Conservation and Power, House Committee on Energy and Commerce, 98 Cong. 1 sess., Sept. 23, 1983, p.450.

4. Harold Denton, memorandum to the Commission, July 23, 1980, p.5.

5. Ibid., p.4.

6. Louise Powell, NRC, note to William Massar, NRC, "Grand Gulf Nuclear Station—Final Supplement to SER," July 9, 1974.

7. James Felton, NRC, letter to Daniel Ford, UCS, FOIA-77-15, Feb. 10, 1977, p.2 and Appendix B.

8. Victor Stello, Jr., memorandum to Voss Moore, NRC, July 25, 1974.

9. Ibid., enclosure, p.25.

10. R. C. DeYoung, memorandum to Robert Heineman, NRC, "Post-CP Evaluations for Grand Gulf," Aug. 4, 1975.

11. Ibid.

12. Grand Gulf Nuclear Station Units 1 and 2, Preliminary Safety Evaluation Report, Amendment 22; see also Walter Butler, NRC, undated letter to MP&L Vice-President N. Stampley. (This letter was neither dated nor mailed but was released in response to UCS FOIA request 77–2.)

13. William Dircks, memorandum to Commissioner Gilinsky, Jan. 13, 1984, p.2.

14. Commission meeting transcript, Briefing on Grand Gulf Technical Specification Requirements, March 20, 1984, pp.9, 12.

15. Thomas Roberts, acting NRC chairman, letter to Rep. Edward Markey, July 17, 1984, enclosure, Answers to Questions 1(F) and 6(C).

16. Rep. Edward Markey, opening statement, Oversight Hearing on the Grand Gulf Nuclear Powerplant, unbound transcript, Subcommittee on Energy and the Environment, House Commission on Interior and Insular Affairs, 98 Cong. 2 sess., July 24, 1984. (Hereinafter, Markey opening statement.)

17. Commission meeting transcript, Discussion and Vote on Full Power Operating License for Grand Gulf, July 31, 1984, p.12.

18. Markey, letter to Chairman Palladino, Aug. 24, 1984.

19. Markey opening statement.

20. General Electric Co. (Vallecitos Nuclear Center—General Electric Test Reactor, Operating License No. TR-1), LBP-82-64, 16 NRC 596, 700 (1982).

21. Commonwealth Edison Co. (Byron Nuclear Power Station, Units 1 & 2), LBP-84-2, 19 NRC 36, 208 (1984).

22. Ibid.

23. Ibid., pp.209–210.

24. Ibid., pp.209–212.

25. Commissioner Gilinsky, testimony, *Nuclear Regulatory Commission Budget Request for Fiscal Years 1984 and 1985*, hearings, Subcommittee on Energy and the Environment, House Committee on Interior and Insular Affairs, 98 Cong. 2 sess., Feb. 9, 1984, p.566.

26. Ibid.

27. Memorandum and Order, Docket No. 50–142 OL, Feb. 24, 1984, p.3.

28. Ibid., pp.6–7.

29. Ibid., p.7, original emphasis.

30. Memorandum and Order, Docket No. 50–142 OL, April 13, 1984, p.13, note 2.

31. James Miller, NRC, affidavit, April 8, 1981, p.2, attached to NRC Staff Motion for Summary Disposition, Docket No. 50–142, April 13, 1981.

32. 10 CFR 73.60.

33. The Regents of the University of California (UCLA Research Reactor), LBP-84-29, 20 NRC 133, 137 (1984).

34. Ibid., pp.137–142.

35. Ibid., p.147.

36. Ibid., pp.147–148.

37. Donald Carlson, NRC, affidavit, April 7, 1981, p.4, note 1, attached to NRC Staff Motion for Summary Disposition, Docket No. 50–142, April 13, 1981.

38. The Regents of the University of California (UCLA Research Reactor), LBP-84-29, 20 NRC 133, 143–145 (1984).

39. Ibid., p.145.

40. Ibid., p.149.

41. Ibid., p.148.

42. Ibid.

43. Ibid., p.147.

44. Ibid., p.148.

45. Subcommittee staff, memorandum to subcommittee members, Subcommittee on Oversight and Investigations, House Committee on Interior and Insular Affairs, 98 Cong. 1 sess., June 17, 1983, enclosure, p.35. (Hereinafter, subcommittee memorandum.)

46. NUREG-0600, draft, July 1979, p.57, B&W exhibit 727.

47. Ibid., section 4.3.2.

48. Robin Clark, "Failed Rancho Seco valve criticized," *San Francisco Examiner*, Oct. 2, 1983, p.B-1.

49. Subcommittee memorandum, enclosure, p.1.

50. John Warshow and Alan MacRobert, "Faulty Nuclear Pumps," *Pacific Sun*, reprint of Dec. 18–24, 1981.

51. *Nuclear Regulatory Commission's Inspection Process: Hayward Tyler Pump Company*, hearing, Subcommittee on Oversight and Investigations, House Committee on Interior and Insular Affairs, 97 Cong. 2 sess., April 6, 1982, pp.22–23. (Hereinafter, Hayward Tyler hearing.)

52. Subcommittee memorandum, enclosure, p.2.

53. David Gamble, Mark Resner, and John Sinclair, NRC, memorandum to James Cummings, April 20, 1982, p.7. (Hereinafter, Gamble-Resner-Sinclair memorandum.)

54. James Cummings, memorandum to Chairman Palladino, March 30, 1982 (hereinafter, Cummings memorandum); Gamble-Resner-Sinclair memorandum, pp.7–8.

55. John Collins, NRC, memorandum to William Dircks, March 5, 1982.

56. Gamble-Resner-Sinclair memorandum, pp.2–5.

57. Ibid., p.5.

58. Ibid., pp.6–7.

59. Cummings memorandum.

60. Nunzio Palladino, testimony, Hayward Tyler hearing, p.88.

61. Ibid., p.7; Gamble-Resner-Sinclair memorandum, pp.7–8.

62. Palladino, testimony, Hayward Tyler hearing, pp.88–89.

63. Cummings memorandum, p.1.

64. Gamble-Resner-Sinclair memorandum, p.1.

65. Cummings, memorandum to David [Gamble], Mark [Resner], and John [Sinclair], April 23, 1982, emphasis added.

66. Hayward Tyler hearing, p.21; see also subcommittee memorandum, enclosure, p.5.

67. Chairman Palladino, letter to Rep. Edward Markey, Jan. 4, 1983, enclosure, Answer to Question 1.

68. NRC Inspection/Investigation Reports 99900345/82–01, 82–02, 82–03, 82–04, as cited in subcommittee memorandum, p.7.

69. Hayward Tyler hearing, pp.7, 21.

70. Subcommittee memorandum, enclosure, p.14.

71. Office of Inspector and Auditor, Report of Investigation, Paul Narbut; Release of Draft Report to Licensee, June 15, 1982, Summary; William Dircks, memorandum to Richard DeYoung et al., March 24, 1982, original emphasis.

72. Office of Inspector and Auditor, Report of Investigation, Paul Narbut; Release of Draft Report to Licensee, June 15, 1982, Summary; R. H. Engelken, NRC, memorandum to J. L. Crews, G. S. Spencer, and A. D. Johnson, NRC, April 9, 1982.

73. Subcommittee memorandum, enclosure, p.10.

74. Markey, letter to Palladino, Feb. 18, 1983.

75. Richard DeYoung, memoranda to Thomas Rehm, NRC, March 2, 1983, and March 10, 1983, as cited in subcommittee memorandum, enclosure, pp.10–22.

76. Palladino, letter to Markey, March 24, 1983, as cited in subcommittee memorandum, enclosure, p.11.

77. Subcommittee memorandum, enclosure, p.11.

78. DeYoung, memorandum to Rehm, March 10, 1983, as cited in subcommittee memorandum, enclosure, p.11.

79. David Gamble, memorandum to James Cummings, June 22, 1982; subcommittee memorandum, enclosure, p.15.

80. Inspection and Enforcement Manual, Chapter 1000, section 1025–04.

81. Subcommittee memorandum, enclosure, pp.15–17.

82. Gamble, memorandum to Cummings, June 22, 1982.

83. Ibid.

84. Cummings, memorandum to the Commission, June 17, 1982, emphasis added.

85. Gamble, memorandum to Cummings, June 22, 1982.

86. John Martin, NRC, memorandum to William Dircks, April 20, 1983, as cited in subcommittee memorandum, enclosure, p.12.

87. Rehm, memorandum to Cummings, May 4, 1983, as cited in subcommittee memorandum, enclosure, p.12.

88. Markey, letter to Palladino, Sept. 21, 1983.

89. Darrell Eisenhut, letter to George Maneatis, Pacific Gas & Electric Co., Sept. 2, 1983, as cited in Markey, letter to Palladino, Sept. 21, 1983.

90. Palladino, letter to Markey, Jan. 9, 1984, enclosure, pp.2–3.

91. Palladino, letter to Markey, Jan. 9, 1984, enclosure, p.3.

92. Markey, letter to Palladino, Nov. 1, 1983.

93. Markey, letter to Palladino, Sept. 21, 1983.

94. Palladino, letter to Markey, Oct. 7, 1983.

95. Deposition of Robert Taylor, NRC, Docket No. 50–445, July 17, 1984.

96. Anthony Roisman, Esq., letter to Samuel Chilk, NRC, Aug. 1, 1984.

97. *Safety Evaluation Report Related to the Operation of Diablo Canyon Nuclear Power Plant, Units 1 and 2*, NUREG-0675, Supplement No. 22, March 1984, pp.A.4-132, A.4-133.

98. J. O. Schuyler, Pacific Gas & Electric Co., letter to Harold Denton, April 30, 1984, attachment, p. 21; D. A. Brand, PG&E, letter to Denton, May 17, 1984, enclosure, pp.4, 9.

99. Subcommittee memorandum, enclosure, pp.19, 22.

100. Lawrence Lippe, DOJ, letter to William Dircks, March 7, 1980, p.3.

101. U.S. Attorney for the District of New Jersey, letter to James Cummings, Aug. 4, 1981, as cited in subcommittee memorandum, enclosure, pp.22–23.

102. Lawrence Lippe, DOJ, letter to James Sniezek, NRC, March 7, 1980, as cited in subcommittee memorandum, enclosure, p.20.

103. Julian Greenspun, DOJ, letter to Ben Hayes, NRC, March 25, 1983, p.3.

104. Ibid., pp.3–4.

105. Ibid., pp.4–5.

106. Ibid., p.5.

107. *TMI-1 Restart: An Evaluation of the Licensee's Management Integrity as It Affects Restart of Three Mile Island Nuclear Station Unit 1*, NUREG-0680, Supp. No. 5, pp.5-1–5-3, 5-6–5-7, July 1984. (Hereinafter, NUREG-0680, Supp. 5.)

108. Ibid., pp.8-16–8-21.

109. U.S. Attorney David Queen, Statement of Facts Submitted by the United States, *U.S. v. Metropolitan Edison Co.*, Feb. 28, 1984, pp.11–14.

110. Ibid., pp.13–17.

111. IE interview of Harold Hartman, May 22, 1979, p.15.

112. Chairman Palladino, letter to Rep. Morris Udall, Dec. 30, 1983.

113. D. Lowell Jensen, DOJ, letter to Palladino, May 17, 1983.

114. Victoria Toensing, DOJ, letter to Udall, Jan. 2, 1985.

115. Commission meeting transcript, Briefing on Staff Revalidation of Management Competence of TMI, May 24, 1983, pp.14–16.

116. Metropolitan Edison Co. (Three Mile Island Nuclear Station, Unit 1), LBP-81-32, 14 NRC 381 (1981).

117. Ibid.

118. Victor Stello, Jr., memorandum to files, June 15, 1983.

119. Commissioner Gilinsky, letter to Udall, Nov. 4, 1983; Peter Bradford, State of Maine Public Utilities Commission, letter to Stello, Nov. 10, 1983.

120. William Dircks, memorandum to the Commission, April 26, 1983.

121. Metropolitan Edison Co. (Three Mile Island Nuclear Station, Unit 1), ALAB-738, 18 NRC 177 (1983); Commission Order, Docket No. 50–289, Oct. 7, 1983.

122. NUREG-0680, Supp. No. 5, pp.5-2–5-6, July 1984.

123. Ibid., p.13-1.

124. *U.S. v. Metropolitan Edison Co.* (M.D. Pa.), Criminal Docket No. 83–00188, Transcript of Proceedings, p.63.

125. *Three Mile Island Alert, Inc. v. NRC.*, 3d Circuit, 771 F.2d 720, 749–750, 1984.

126. Julian Greenspun, DOJ, letter to Ben Hayes, NRC, March 23, 1984.

127. Michelle Adato, telephone conversation with Roger Fortuna, NRC, Jan. 17, 1985.

128. Adato, telephone conversation with Ed Fay, NRC, Jan. 17, 1985.

129. Ibid.

130. Ibid.

131. Ben Hayes, memorandum to Chairman Palladino, enclosure, "Report of Investigation, Three Mile Island Nuclear Generating Station, Unit 2 Allegations Regarding Safety Related Modification, Quality Assurance Procedures and Use of Polar Crane, Case No. H-83-002," Sept. 1, 1983.

132. "Staff Review and Response to OI Report on TMI-2 Clean-up Allegations," SECY-84-36, Jan. 25, 1984; "NRC Staff Decides OI Was Right After TMI-2 Polar Crane Brake Failures," *Inside NRC*, Nov. 12, 1984, p.6.

133. William Dircks, memorandum to the Commission, Oct. 29, 1984, as cited in *Inside NRC*, Nov. 12, 1984, p.6.

134. H. Rept. 98–217, 98 Cong. 1 sess., p.138.

135. *Inspection Report of "Preliminary Report, Seismic Design Reverification Program" at Diablo Canyon Nuclear Power Plants, Units 1 & 2*, NUREG-0862, Issue 2, Jan. 18, 1982, pp.11–12. (Hereinafter, NUREG-0862.)

136. Ibid., pp.17–18.

137. Ibid., p.19; Rep. Morris Udall, letter to Chairman Palladino, April 26, 1982, enclosure, committee staff background paper, p.1. (Hereinafter, background paper.)

138. Background paper, p.13.

139. NUREG-0862, pp.6–7.

140. Background paper, p.31.

141. Subcommittee memorandum, enclosure, pp.21–31.

142. Interview of Barcley Lew, NUREG-0862, Appendix E, pp.478–479.

143. Background paper, pp.23–24.

144. Ibid., pp.27–28.

145. Ibid., p.28.

146. Pacific Gas and Electric Co. (Diablo Canyon Nuclear Power Plant, Units 1 & 2), CLI-82-1, 15 NRC 225 (1982).

147. Background paper, pp.2, 31.

148. Pacific Gas and Electric Co. (Diablo Canyon Nuclear Power Plant, Units 1 & 2), CLI-82-1, 15 NRC 225 (1982), "Additional Views of Commissioner Gilinsky Regarding Pacific Gas and Electric's Material False Statement." (Hereinafter, "Additional Views.")

149. Rep. Richard Ottinger, letter to Chairman Palladino, April 9, 1982.

150. "Additional Views"; Rep. John Dingell and Rep. Ottinger, letter to Palladino, Feb. 23, 1982.

151. Rep. Morris Udall, letter to Palladino, Feb. 8, 1982.

152. James Cummings, memorandum to Palladino, "Subject: Review of Investigation Reported in NUREG-0862," July 30, 1982.

153. Ibid.

154. Ibid.

155. Subcommittee memorandum, enclosure, pp. 30–31.

156. Robert Cloud et al., Testimony on Behalf of the Independent Design Verification Program Regarding Contentions 1, 2, and 5–8, pp. 1/2–2 and Attachment A to Independent Design Verification Program Exhibit List.

157. Report by majority staff, Committee on Interior and Insular Affairs,

Reporting of Information Concerning the Accident at Three Mile Island, 97 Cong. 1 sess., March 1981.

158. Commissioner Gilinsky's Separate Remarks, attached to letter from Palladino to Udall, March 2, 1982.

159. *Investigation into Information Flow During the Accident at Three Mile Island*, NUREG-0760, January 1981, p.11.

160. David Gamble, prefiled testimony, In the Matter of Metropolitan Edison Co. (Three Mile Island Nuclear Station, Unit No. 1), Docket No. 50-289, Nov. 5, 1984, p.3. (Hereinafter, Gamble testimony.)

161. Ibid., p.4.

162. Ibid., pp.5–6.

163. Ibid., p.6.

164. Ibid.

165. David Gamble, NRC, memorandum to Norman Moseley, NRC, Jan. 26, 1981.

166. Ibid.

167. Gamble testimony, p.7.

168. Udall, letter to Palladino, Feb. 4, 1982.

169. Henry Myers, Committee on Interior and Insular Affairs, letter to William Dircks, Norm Haller, March 8, 1982.

170. Commission meeting transcript, Briefing on Information Flow Concerning the TMI Accident, Jan. 22, 1981, pp.14–20.

171. Commissioner Gilinsky's Separate Remarks, attached to letter from Palladino to Udall, March 2, 1982.

172. Udall, letter to Palladino, Feb. 4, 1982.

173. Steve Stecklow, "Death of a nuclear plant: The regulatory breakdown," *Philadelphia Inquirer*, reprint of Jan. 24–25, 1984, p.13. (Hereinafter, Stecklow.)

174. Ibid.

175. Interview of Terry Harpster, conducted by OIA investigators David Gamble and John Sinclair, March 6, 1981, attached to James Cummings, memorandum to Leonard Bickwit, NRC, Jan. 5, 1983. (Hereinafter, Harpster interview.)

176. Stecklow.

177. *Report to the Chairman on an Investigation into Allegations of Thomas Applegate Concerning the Conduct of the Office of Inspector and Audit*, conducted by Judge Helen Hoyt, ASLB, and C. Sebastian Aloot, OGC, July 12, 1983, pp.7–10. (Hereinafter, Hoyt report.)

178. IE Investigation Report No. 50-358/80-09, July 2, 1980.

179. Hoyt report, p.6.

180. James Cummings, OIA, memorandum to the Commission, enclosure, "Special Inquiry Re: Adequacy of IE Investigation 50-358/80-09 at the William H. Zimmer Nuclear Power Station," Aug. 7, 1981.

181. Ibid., pp.1–2.

182. Interview of James McCarten, conducted by Judge Helen Hoyt and C. Sebastian Aloot, June 7, 1983 (attached to Hoyt report), p.20. (Hereinafter, McCarten interview.)

183. Stecklow.

184. McCarten interview, p.22.

185. Stecklow.

186. Geraldine Brooks, "James Keppler, Chief of NRC in Midwest, is Beset by Problems," *Wall Street Journal*, Aug. 28, 1984, p.1. (Hereinafter, Brooks.)

187. *Quality Assurance at the Midland Powerplant*, hearing, Subcommittee on Energy and the Environment, House Committee on Interior and Insular Affairs, 98 Cong. 1 sess., June 16, 1983, p.52.

188. Brooks.

189. Ibid.

190. Ibid.

191. Stecklow; McCarten interview, p.100.

192. McCarten interview, p.101; Stecklow.

193. Stecklow.

194. Hoyt report, pp.13–14.

195. Ibid., p.30, original emphasis.

196. Victor Stello, Jr., memorandum to Chairman Palladino, Aug. 31, 1983, pp.4–5.

197. Julian Greenspun, DOJ, letter to Ben Hayes, NRC, March 25, 1983.

198. Hoyt report, p.14.

199. Investigation Report No. 50-358/81-13, attachment to letter from James Keppler, NRC, to W. R. Dickhoner, Cincinnati Gas & Electric, Nov. 24, 1981, pp.6–7.

200. McCarten interview, pp.74–76.

201. Ibid., p.77.

202. James Keppler, memorandum to the Commission, Sept. 21, 1983.

203. McCarten interview, pp.133–34.

204. Hoyt report, p.19.

205. Interview of Anne Tracy, conducted by Judge Helen Hoyt, June 8, 1983 (attached to Hoyt report), pp.49–51.

206. Hoyt report, pp.24–25.

207. James Cummings, memorandum to Leonard Bickwit, Jan. 5, 1983, p.2.

208. Harpster interview.

209. Hoyt report, pp.26–27.

210. Ibid., p.28, emphasis added.

211. Ibid., pp.28–29.

212. *Applegate v. NRC*, U.S.D.C., D.D.C. No. 82-1829, May 24, 1983.

213. Rep. Morris Udall, letter to Chairman Palladino, Dec. 7, 1982.

214. Palladino, letter to Udall, April 14, 1983, pp.1–4.

215. Palladino, letter to Udall, April 14, 1983, pp.3–5.

216. NRC Announcement No. 12, Feb. 8, 1984.

217. *Applegate v. NRC*, U.S.D.C., D.D.C. No. 82-1829, May 24, 1983.

218. Palladino, memorandum to Stello, Oct. 27, 1983.

219. Julian Greenspun, DOJ, letter to Ben Hayes, NRC, March 25, 1983.

220. Daniel Donoghue, NRC, memorandum to all SES members, Sept. 30, 1981; Patricia Norry, NRC, memorandum to all SES members, Oct. 6, 1982; NRC Announcement No. 102, Oct. 4, 1983.

221. Letter from Markey to Palladino, Dec. 6, 1983, and Palladino response, Dec. 21, 1983.

222. James Cummings, memorandum to William Dircks, Oct. 31, 1981, as cited in subcommittee memorandum, enclosure, p.28.

223. Samuel Chilk, NRC, memorandum to Cummings, Nov. 10, 1981, as cited in subcommittee memorandum, enclosure, p.28.

224. Cummings, memorandum to the Commission, Dec. 14, 1981, as cited in subcommittee memorandum, enclosure, pp.28–29.

225. Subcommittee memorandum, p.29.

226. See San Luis Obispo Mothers for Peace, petitions (and supplements) to the NRC pursuant to 10 CFR 2.206, Docket No. 50-275, Feb. 2, 1984; March 1, 1984; March 23, 1983; April 12, 1984; May 3, 1984; June 21, 1984; and July 16, 1984.

227. Commission meeting transcript, Discussion of Pending Investigation at Diablo Canyon, Jan. 23, 1984, pp.27–28.

228. Chairman Palladino, memorandum to the commissioners, March 20, 1984.

229. "Late Allegations," SECY-84-249/A, Aug. 31, 1984, p.1.

230. Commission meeting transcript, Comments by Parties on Diablo Canyon Criticality and Low Power Operation, Feb. 10, 1984, pp.73, 76.

231. Commission meeting transcript, Discussion/Possible Vote on Diablo Canyon Criticality and Low Power Operation, April 13, 1984, pp.44–48 (hereinafter, Diablo Canyon transcript); *Safety Evaluation Report Related to Operation of Diablo Canyon Nuclear Power Plants Units 1 & 2*, NUREG-0675, Supp. 22, March 1984, p.E-4.

232. Diablo Canyon transcript, pp.86–87.

233. Deposition of Isa Yin, read into Diablo Canyon transcript, p.82.

234. Affidavit of Thomas Devine, Government Accountability Project, Docket No. 50-275, July 11, 1984, p.2. (Hereinafter, Devine affidavit.)

235. Ibid., pp.2–4.

236. Statement of Isa Yin, in *Licensing Process at Diablo Canyon Nuclear Powerplant*, hearing, Subcommittee on Energy and the Environment, House Committee on Interior and Insular Affairs, 98 Cong. 1 sess., Aug. 30, 1984, p.131. (Hereinafter, Diablo Canyon hearing.)

237. Devine affidavit, p.3.

238. Diablo Canyon hearing, pp.16–17.

239. Ibid., p.17.

240. Ibid., pp.63–64.

241. Devine affidavit, p.3.

242. 291st Meeting of the ACRS, transcript, July 13, 1984, p.434.

243. Diablo Canyon hearing, p.10.

244. "NRC Engineer Resigns from Diablo Task Force," *Washington Post*, July 12, 1984.

245. Devine affidavit, p.4.

246. Diablo Canyon hearing, p.17, emphasis added.

247. Pacific Gas and Electric Co. (Diablo Canyon Nuclear Power Plant, Units 1 & 2), CLI-84-13, 20 NRC 267, 281-282 (1984), "Dissenting Views of Commissioner Asselstine."

248. *San Luis Obispo Mothers for Peace v. United States Nuclear Regulatory Commission*, D.C. Circuit, No. 84-1410, Aug. 17, 1984.

249. Statement of Commissioner Asselstine, in Diablo Canyon hearing, pp.129–130; Representatives Jerry Patterson and Leon Panetta, letter to the Commission, Oct. 5, 1984.

250. Diablo Canyon hearing, pp.19–20.

251. Ibid., p.20.

6. Conclusions and Recommendations

1. Chairman Palladino, letter to Rep. Thomas Bevill, Jan. 23, 1985, p.11.

2. *Precursors to Potential Severe Core Damage Accidents: 1969–1979, A Status*

Report, NUREG/CR-2497, Vol.1, ORNL/NSIC-182/VI; *Precursors to Potential Severe Core Damage Accidents: 1980–1981, A Status Report*, NUREG/CR-3591, Vol.1, ORNL/NSIC-217/VI.

3. John Kemeny et al., *Report of the President's Commission on the Accident at Three Mile Island*, October 1979, p.62.

4. Mitchell Rogovin et al., *three mile island: A Report to the Commissioners and to the Public*, January 1980, pp.142–144, emphasis in original.

INDEX